MATHEMATICAL CIRCLES SQUARED

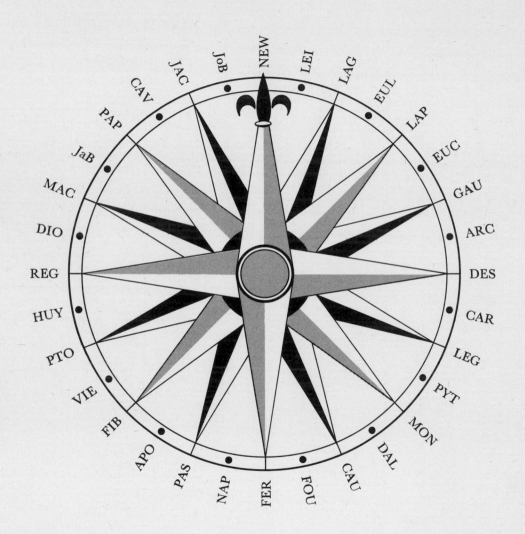

MATHEMATICAL CIRCLES SQUARED

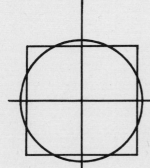

A THIRD COLLECTION OF
MATHEMATICAL STORIES AND ANECDOTES

HOWARD W. EVES

PRINDLE, WEBER & SCHMIDT, INC.

Boston, Massachusetts

FRONTISPIECE: "Boxing the mathematical compass" is described in Item 315°.

Library of Congress Catalog Card Number: 76-175520
SBN 87150-154-6

Printed in the United States of America

TO   VERNER E. HOGGATT, JR.

who, over the years, has sent me more
mathematical goodies than anyone else

PREFACE

Here is a third trip around the mathematical circle, and, at least for some time, it must be the last such trip. I cannot deny that it has been fun making these trips, but there are so many other, and more serious, things I wish to write that it is high time I set the mathematical stories and anecdotes aside—with somewhat more than half of my collection now in print.

Except for the first quadrant and the tail end of the fourth quadrant, the items in this third set of stories are classified by geography, in contrast to the classification by chronology in *In Mathematical Circles* and by subject matter in *Mathematical Circles Revisited*.

At one time it had been my intention to devote a whole and separate work to the subject of mathematically motivated designs. With more pressing writing commitments on hand, I have decided to forego this project, and have incorporated a couple of dozen examples in the first quadrant of the present work. Perhaps some reader may be inspired to take up the project and carry it out more fully and more deservingly.

Once again, I express thanks to those colleagues who have shown interest in my trips about the mathematical circle and who have sent me favorite items for possible inclusion. It has been my desire to acknowledge these contributors, and any other special sources, at the end of the items concerned; I hope I will be forgiven if I have unintentionally overlooked anyone. Special thanks go to Academic Press and to Dover Publications for permitting me to quote from some of their books.

As with the preceding trip around the circle, this third one was written intermittently during a (second) year spent assisting at the Gorham unit of the University of Maine. Upon completion of this second year, I return to the central campus at Orono, carrying with me exceedingly pleasant memories of my two enjoyable years at Gorham.

HOWARD W. EVES

CONTENTS

PREFACE *ix*

QUADRANT ONE

NUMBERS AND NUMBER RECKONING

1° Saint Jerome's commentary on a passage in Matthew, *3*
2° The ring finger, *4*
3° Finger counting in Rome, *4*
4° A reversal, *5*
5° Richer or poorer?, *5*
6° Holding the bowstring, *5*
7° Attitude for prayer, *5*
8° Frederick II on falconry, *5*
9° Fibonacci and finger counting, *6*
10° The Pa-kua, *6*
11° One way to put it, *7*
12° A misplaced credit, *7*
13° The origin of the printers' ellipsis, *7*
14° Wise men versus fools, *8*
15° An unintentional commentary, *8*
16° An old order giving way, *8*
17° The case for the abacists, *8*
18° How Xerxes counted his army, *9*
19° Popular immensities, *9*
20° Overestimation, *10*
21° Perfect numbers, *10*
22° You can't fool the sergeant, *10*
23° An application of the "least natural number principle," *10*

xi

Contents

24° Lehmer's table of primes, *11*
25° Modern numerals, *11*
26° Chalk one up for the human being, *12*
27° Broadening the college language requirement, *13*
28° A possible hope for the world, *13*
29° A proud boast, *13*

Magic Squares

30° Some terminology and a celebrated unsolved problem, *14*
31° The lo-shu, *15*
32° The early Hebrews and the lo-shu, *16*
33° Talismans and charms, *16*
34° The earliest fourth-order magic square, *18*
35° Albrecht Dürer's magic square, *18*
36° A pandiagonal magic square on a torus, *20*
37° A pandiagonal magic square on a plane, *20*
38° A pandiagonal magic square on a hypercube, *21*
39° A magic square in architectural decoration, *22*
40° Magic squares of odd order, *22*
41° Magic squares of doubly-even order, *23*
42° Doubly-magic and trebly-magic squares, *25*
43° Prime magic squares, *26*
44° Fibonacci magic squares, *26*
45° Magic cubes, *27*
46° Brains versus machines, *28*
47° Origin of Eulerian squares, *30*

Mathematically Motivated Designs

48° Some artistic applications of the lo-shu, *31*
49° Magic traces, *31*
50° Re-entrant knight's path, *33*
51° Half-board solutions, *35*
52° Magic re-entrant half-board knight's paths, *36*
53° Re-entrant strong-knight paths, *36*
54° Magic re-entrant king's path, *37*
55° A conversation piece, *37*
56° A patio mosaic, *38*
57° The octagon puzzle, *39*
58° A toral tulip garden, *41*
59° A flower garden on a projective plane, *41*

Contents

60°   Schlegel flower gardens,   *41*
61°   Professor Tucker's flag,   *43*
62°   More patio designs,   *43*
63°   In the foyer of the mathematics building,   *45*
64°   Room numbers,   *45*
65°   Tying gift boxes,   *46*
66°   A Pythagorean theorem design,   *46*
67°   A maple-leaf tile,   *47*
68°   Another maple-leaf tile,   *48*
69°   Some other tiles,   *48*
70°   A lamp base,   *49*
71°   A quilt,   *49*
72°   Some more quilts,   *52*

Geometry

73°   An analogy,   *56*
74°   Another analogy,   *56*
75°   A science fiction possibility,   *56*
76°   The truth of a theory,   *56*
77°   Zeuxis and Apollodorus,   *57*
78°   Maps,   *58*
79°   An anthology of great poetry,   *59*
80°   The method of dual languages,   *59*
81°   Blind geometers,   *59*
82°   The true meaning of the Delphian oracle,   *60*
83°   The origin of geometry,   *60*
84°   Geometers and analysts,   *61*
85°   DIN,   *61*
86°   An early love affair,   *62*
87°   Albrecht Dürer's approximate constructions,   *62*
88°   Two daffynitions,   *64*
89°   The birthplace of topology,   *64*
90°   Q. E. D.,   *66*

QUADRANT TWO

From the American Scene

91°   Super-respectability,   *69*

xiii

CONTENTS

92°   Practical rambler, *69*
93°   Mountain climber, *70*
94°   Creator of suspense, *70*
95°   Epicurean, *70*
96°   Einstein and religion, *70*
97°   A sufficient reward, *70*
98°   Einstein and his check, *71*
99°   The prankster, *71*
100°   Identification, *71*
101°   How Norbert Wiener got his beard, *71*
102°   Wienerian wit, *72*
103°   The modern physicist, *72*
104°   Shop-talker, *72*
105°   Wienerian disease, *72*
106°   An incomplete story, *72*
107°   How a young mathematician saved himself by ingenuity, *72*
108°   Crackpots, *73*
109°   Benjamin Franklin Finkel, *73*
110°   Problemist extraordinaire, *74*
111°   Two prizes, *75*
112°   A student from Missouri, *75*
113°   Cupid's problem, *75*

AMONG THE ENGLISH
114°   Blissard's sermon, *76*
115°   From Blissard's obituary, *77*
116°   Unattached, *77*
117°   A man of peculiarities, *78*
118°   Mrs. De Morgan, *78*
119°   Definite integrals, *79*
120°   No ear for music, *80*
121°   A legend, *80*
122°   Mathematical talent foreshadowed, *80*
123°   Kite flying, *80*
124°   Unlocking a puzzle, *81*
125°   A repeated tribute, *81*
126°   Eddington on the fourth dimension, *81*
127°   Student, *82*
128°   Preparing for an examination, *82*
129°   Littlewood and the Dedekind cut, *82*

CONTENTS

130°   Littlewood and Hardy, *83*
131°   Littlewood and Veblen, *83*
132°   A comment by J. E. Littlewood, *83*
133°   Littlewood demonstrates mountain-climbing techniques, *83*

Two Irishmen
134°   Poetry and disappointment, *84*
135°   Hamilton and Wordsworth, *86*
136°   A rare compliment, *87*
137°   Hamilton's second love affair, *87*
138°   A bad choice, and alcohol, *88*
139°   A rare honor, *88*
140°   Hamilton's wish, *89*
141°   Modified Irish blarney, *89*
142°   On the turn of a penny, *89*
143°   His last act, *90*

Two Scotsmen
144°   Wiping away a stigma, *91*
145°   *T and T*, *91*
146°   Tait and Stevenson, *91*
147°   Holding back the crowd, *91*
148°   Keeping the demonstration desk clear, *91*
149°   Behind the times, *92*
150°   The Tait's compass, *92*
151°   The hare and the hounds, *93*
152°   An embarrassment, *93*
153°   An allegory, *93*
154°   Unpopular, *94*
155°   Taking roll, *94*
156°   Poor arithmetician, *94*
157°   The marble, *94*

The Last Universalist
158°   Poor motor coordination, *95*
159°   The artist, *95*
160°   Fame helps, *95*
161°   How does he do it?, *95*
162°   A misunderstanding, *96*
163°   Cousins, *96*

164° Literary success, *96*
165° A poor administrator, *97*
166° A powerful and unusual memory, *97*
167° Poincaré's manner of working and composing mathematics, *97*
168° Rated an imbecile, *98*
169° Two kinds of absentmindedness, *98*
170° A lover of animals, *99*
171° An act of heroism, *99*
172° Sylvester meets Poincaré, *99*
173° An unfinished symphony, *99*

CROUTONS FOR THE FRENCH SOUP
174° Deep insight, *100*
175° Precision, *100*
176° Jordan and notation, *100*
177° A story of Hadamard's youth, *101*
178° A story of Hadamard's older age, *101*
179° Bourbaki's office, *101*
180° Mathematicians are like Frenchmen, *102*

QUADRANT THREE

TWO NORWEGIANS AND A RUSSIAN
181° A lost manuscript, *105*
182° By studying the masters, *106*
183° Spurred on because of a bully, *106*
184° An unfortunate incident, *106*
185° An almost unique event, *107*
186° A most admirable trait, *107*
187° An apt description, *108*

THE PRINCE OF MATHEMATICIANS
188° A student's lamp, *108*
189° Lost fame, *108*
190° Ribbentrop as an overnight guest, *109*
191° Ribbentrop as an observatory guest, *109*
192° Gauss's assistant, *110*
193° Gauss's power of concentration, *110*

Contents

194°   Source of pleasure, *110*

195°   Maintaining a standard, *110*

196°   Some similarities between father and son, *111*

197°   Gauss and his grandson, *111*

198°   Gauss and the British Admiralty, *111*

199°   The Berlin Academy of Sciences, *112*

200°   Wrong order, *112*

201°   Theory and practice, *112*

202°   A sacrilege, *113*

203°   The hunt rather than the treasure, *113*

204°   Gauss and a student, *113*

205°   The onerous side of teaching, *113*

206°   Cayley on Gauss, *114*

207°   Absolute and relative properties of a surface, *114*

208°   The heliotrope, *114*

209°   The Hohenhagen Tower, *115*

210°   The Gauss-Weber monument, *115*

211°   Ravages of war, *115*

Three Great Göttingen Professors

212°   An interesting analogy. *116*

213°   Presents, *116*

214°   Rapid assimilation, *116*

215°   Contrast, *116*

216°   Riemann's Habilitationsschrift, *117*

217°   Genius and craftsmen, *117*

218°   Intellectual tonnage, *118*

219°   A prodigy, *118*

220°   At a department meeting, *119*

221°   A rare burst of arrogance, *119*

222°   Minkowski on Einstein, *119*

223°   Minkowski on Mrs. Hilbert, *119*

224°   Minkowski's death, *120*

225°   The spoiled child of mathematics, *120*

226°   A Landau game, *120*

227°   The two sides of the ledger, *121*

228°   Felix and Serge, *121*

229°   Fermat's last theorem, *121*

230°   Irrepressible, *121*

231°   Slightly incorrect, *122*

CONTENTS

232° Landau leaves Göttingen, *122*
233° Mrs. Landau, about her husband, *122*

THE MASTER

234° Stimulation, *123*
235° Hilbert conducts business with the Minister of Culture, *124*
236° The Wolfskehl Prize, *124*
237° Inflation, *125*
238° Requirements for a good problem, *125*
239° Catching a fly on the moon, *125*
240° In defense of Galileo, *125*
241° Hilbert's "flames," *125*
242° Brutally direct, *126*
243° Hilbert's harshness, *127*
244° Geheimrat Hilbert, *127*
245° Poor memory and slow understanding, *127*
246° Hilbert's obtuseness, *128*
247° Hilbert's class lectures, *128*
248° Existence, *128*
249° Hilbert and the new quantum mechanics, *128*
250° Hilbert and the prime numbers, *129*
251° Student adulation, *129*
252° Hilbert and Ackermann, *130*
253° The Riemann hypothesis, *130*
254° Hilbert and Hilbert space, *130*
255° Taking credit, *131*
256° Franz Hilbert, *131*
257° Imagination, *132*
258° Point of view, *132*
259° The frog and mouse battle, *132*
260° Hilbert on the law of excluded middle, *133*
261° No roast goose, *133*
262° Memorial to Darboux, *134*
263° Declaration to the Cultural World, *134*
264° A medical success story, *134*
265° An anti-Semite, *135*
266° Bieberbach and the Second International Congress, *135*
267° Hilbert and the Second International Congress, *136*
268° Hilbert's hotel bill, *136*
269° Hilbert Strasse, *136*

Contents

270° A comparison of two deaths, *136*

QUADRANT FOUR

Further Göttingen Mathematicians

271° Ordinary and extraordinary, *141*
272° Felix the Great, *141*
273° A distant god, *141*
274° A faux pas, *141*
275° A syllogism, *141*
276° Runge as a calculator, *142*
277° Runge introduces skiing, *142*
278° A proof of an impossibility, *143*
279° Walzermelodie, *143*
280° No effort too great, *143*
281° Blumenthal's dissertation, *143*
282° Blumenthal's end, *143*
283° First member of the honors class, *144*
284° Emmy Noether as a lecturer, *144*
285° The Noether boys, *144*
286° Hilbert and Emmy Noether, *144*
287° Max Born's examination, *144*
288° Twice disappointed, *145*
289° The Bohr brothers, *145*
290° Harald Bohr's examination, *145*
291° Siegel goes to Göttingen, *145*
292° A poor prediction, *145*
293° A paper for the *Annalen*, *146*
294° Van der Waerden as a boy, *146*
295° How Jacob Grommer was led to mathematics, *147*

More German Mathematicians

296° Reward for a discovery, *148*
297° Discouragement, *148*
298° A Kummer simile, *148*
299° Kummer unexpectedly wins a prize, *148*
300° Kummer's empathy, *149*
301° Kummer's last nine years, *149*
302° Lotus-eaters, *149*

Contents

303°   Hot from the forge, *149*
304°   Blaschke's bad mark, *150*
305°   Immortality, *150*

Olla-Podrida

306°   A means of escape, *150*
307°   Depressed mathematicians, *150*
308°   The spirit of mathematical inquiry, *150*
309°   The sure test, *151*
310°   A truism, *151*
311°   Pioneer work is clumsy, *152*
312°   How to win a war, *152*
313°   The Greeks versus the European Christians, *152*
314°   The Greeks versus the Romans, *153*
315°   Boxing the mathematical compass, *153*
316°   Internationalism, *153*
317°   The performers and the creators, *153*
318°   Virtuosity versus depth, *154*
319°   A lesson in freedom, *154*
320°   A difference between mathematics and other arts, *154*
321°   The grove of scholarship, *155*
322°   A law of genetics, *155*
323°   Cheating, *155*
324°   Equivalence and identity, *155*
325°   A proposal and its acceptance, *156*
326°   On large numbers, *156*
327°   Dependence, *156*
328°   The Alfonsine Tables, *157*
329°   Bad news, *157*
330°   A consuming proof, *157*
331°   What a sprouting acorn said, *158*
332°   A revolutionary sputnik, *158*
333°   The queen of the sciences, *158*
334°   An accidental immortality, *158*
335°   The Scopes trial, *158*
336°   Hermannus Contractus, *159*

Printers and Books

337°   The copy editor, *160*
338°   A typist's error, *160*

Contents

339° A questionable practice, *160*
340° Linguistic oddities, *160*
341° Dash it all, *160*
342° Proofreading, *161*
343° The compositor, *161*
344° A ghost-author, *161*
345° An unkind textbook reference, *161*
346° Another unkind textbook reference, *162*
347° Weitzenböck's "Invariantentheorie," *162*

Psychology

348° Gauss's experience, *162*
349° Hamilton's experience, *163*
350° Poincaré's experience, *163*
351° Hadamard's experience, *165*
352° Loss of interest, *165*
353° A mathematical dream, *165*
354° The mathematics bump, *166*
355° Servants, not masters, *166*
356° At what age?, *167*
357° Morning or evening?, *167*
358° Preparation-incubation-illumination, *167*
359° Two kinds of discovery, *167*
360° The thought curve, *168*

ADDENDA

Esthetics

1081° The esthetic appreciation of mathematics, *173*
1082° Esthetic appreciation smothered, *174*
1083° The practical versus the esthetical, *174*
1084° Popular appeal of mathematics, *174*
1085° The guide for the choice of research subjects, *175*
1086° Judging a graduate research student, *175*
1087° A case history, *176*
1088° Another case history, *176*
1089° $= [(33)^2]°$ Ten pertinent quotations, *176*

Index, *179*

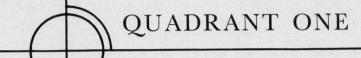

QUADRANT ONE

*From a divine's commentary
to "quite elegantly done"*

NUMBERS AND NUMBER RECKONING

THERE are lots of stories—many of them of pedagogical value—about numbers and number reckoning. Here are some which do not appear in our two earlier collections. Several of these stories refer to the one-time popular finger numbers and to the time when number reckoning was done on a counting board or abacus.

1° *Saint Jerome's commentary on a passage in Matthew.* Saint Jerome (ca. 340–420) was born in a small village in the Illyrian Alps. When about thirty-four, he commenced his five years of hermit life in the desert near Antioch, after which he was ordained and went to Constantinople and began his intensive study of the Scriptures. After a short return to Rome, he settled, in 386, as a recluse and student in a monastery near Bethlehem. His great knowledge of the Scriptures rendered him foremost among the early interpreters of the Bible, and his translation of the Old Testament from Hebrew into Latin became the Vulgate, or popular, version officially used today by the Roman Catholic Church. He also revised the Latin New Testament, referring to the best Greek manuscripts available at his time. In Albrecht Dürer's famous painting, Saint Jerome is pictured in a small house studiously bent over a writing desk and absorbed in his translation of the Bible.

In addition to translating the Bible, Saint Jerome wrote extensive commentaries on the Scriptures, including an especially interesting one on Matthew 13:8, concerning the Parable of the Sower:

> But others fell into good earth, and brought forth fruit, some an hundredfold, some sixtyfold, some thirtyfold.

Here is Saint Jerome's commentary on this passage:

> The hundredfold and the sixtyfold and the thirtyfold fruit, though they may have sprung from the same soil and the same seed, nevertheless differ greatly in their number. Thirty is a symbol of marriage: for this manner of placing the fingers, which are joined and interwoven as in a sweet embrace, represents both man and wife.
>
> Sixty stands for widowhood, because the widow endures trouble and

3

tribulation that weigh on her, just as the thumb is pressed down by the index finger which lies above it. But the more difficult it is to abstain from the pleasures that were once enjoyed, the greater will be the reward.

But now attend carefully, dear Reader: the number hundred is transferred from the left hand to the right, where it is formed with the same fingers but not on the same hand as the hand which symbolized marriage and widowhood.

In order to understand this passage, one must be familiar with the finger counting of medieval times. The number 30 was formed on the left hand by placing the tip of the index finger on the end of the thumb, forming a circle called "the tender embrace," the other fingers of the hand being outstretched. The number 60 was formed on the left hand by pressing the thumb into the palm with the end of the index finger, the other fingers of the hand being outstretched. The various hundreds were formed on the right hand.

Saint Jerome's commentary bears excellent witness to the popular use of finger counting in his day, for his commentary is quite unintelligible to anyone lacking that knowledge.

One still sees, today, "the tender embrace" used as a gesture of something sweet or pleasant or successful.

2° *The ring finger.*   In ancient times it was believed that there is a vein running directly from the heart to the finger next to the little finger. This finger was accordingly called the *medicus*, and, because of its supposed connection with the heart, it was deemed to be the most proper finger to bear any sort of sacred ring. The finger thus became known as the *ring* finger. When, in medieval times, finger counting was developed and systematized, it was decided that 6, the first perfect number (a perfect number is one which is equal to the sum of its proper divisors: $6 = 1 + 2 + 3$), should be formed on the sacred ring finger. It thus turned out that 6 was represented by folding down the ring finger on the left hand.

3° *Finger counting in Rome.*   Pliny the Elder, who died in A.D. 79, has told us of a statue of the two-headed god Janus in the forum at Rome. He said that the statue showed Janus representing the 365 days of the year on his fingers, by forming—according to the common method of finger counting of the time—the number 300 with his right hand and the number 65 with his left.

4° *A reversal.* In Roman finger counting, the numbers less than 100 were formed on the left hand, while the hundreds were formed on the right hand. Among the Arabs, the roles played by the two hands were reversed, in accordance with Arabic script which runs from right to left.

5° *Richer or poorer?* An Arabic poet once cleverly scoffed another Arabic poet, named Khalid, who had become very rich, by the following line: "Khalid went with 90 and came back with 30." It would at first seem that Khalid returned poorer. But the number 90 was formed (by the Arabs) on the fingers of the right hand by laying the index finger on the base of the thumb, thus forming a very small or "lean" circle; the number 30 was formed on the fingers of the right hand by joining the tips of the index finger and thumb, thus forming a large or "rich" circle.

6° *Holding the bowstring.* An Arabic script describes how the bowstring is to be held when shooting an arrow. It is to be grasped with the tips of the index finger and thumb of the right hand, "as the number 30 is formed."

7° *Attitude for prayer.* Another Arabic script describes the proper physical position for prayer by saying that the believer should place his right hand on his leg as though he were forming the number 54 (that is, only the index finger should be extended).

8° *Frederick II on falconry.* Emperor Frederick II (1194–1250) was a noted patron of learning and of the arts. He also enjoyed hunting and wrote a famous book on falconry (*On the Art of Hunting with Birds*). In his book he describes how a huntsman should hold the falcon:

> The hand is held facing neither inward nor outward, but extended in the same direction that the arm is extended. The index finger is then laid over the extended thumb and bent forward over the end segment of the thumb, just as the masters of computation form the number 70

5

with their fingers. The other fingers of the hand are bent over the palm beneath the index finger and thumb, to support the latter, just as the number 3 is formed. Thus the index finger is bent over the thumb and the three other fingers beneath it, in the manner of a master of computation forming the number 73.

9° *Fibonacci and finger counting.*   It is interesting that Fibonacci, who was the great advocate for the use of the Hindu-Arabic numeral system with its place-value notation and convenient computing algorithms, also strongly recommended the retention of finger counting. For in finger counting he found an excellent way for a computer to retain temporary numbers that are to be carried or that are partial results of a more lengthy computation, such as a long division.

10° *The Pa-kua.*   In the ancient Chinese classic, the *I-king*, or *Book of Permutations*, appear the *Liang I*, or "two principles" (the male *yang*, —, and the female *ying*, — —. From these were formed the *Sz' Siang*, or four figures,

and the *Pa-kua*, or eight figures,

The eight symbols of the Pa-kua had various attributes assigned to them (see Figure 1), and hence came to be used in divination.

heaven	stream	fire	thunder	wind	water	mountain	earth
S.	S.E.	E.	N.E.	S.W.	W.	N.W.	N.

FIGURE 1

6

Though there is no supporting historical evidence, one cannot help but see, in the Pa-kua, numerals based on the scale of two. For if we take — as one and — — as zero, the successive trigrams shown in Figure 1, beginning at the right, would represent the numbers 0, 1, 2, 3, 4, 5, 6, and 7. The Pa-kua are today found on diviners' compasses, fans, vases, other household objects, and various kinds of talismans.

11° *One way to put it.*   In the binary system we count on our fists instead of on our fingers.

12° *A misplaced credit.*   The so-called *Attic*, or *Herodian*, Greek numerals were developed some time prior to the third century B.C. and constitute a simple grouping system to base 10 formed from initial letters of number names. In addition to the symbols *I*, Δ, *H*, *X*, *M* for 1, 10, 10^2, 10^3, 10^4, there is a special symbol for 5. This special symbol is an old form of Π, the initial letter of the Greek *pente* (five), and Δ is the initial letter of the Greek *deka* (ten). The other symbols can be similarly explained. The symbol for 5 was frequently used both alone and in combination with other symbols in order to shorten number representations. As an example in this numeral system we have

$$2857 = \text{X X Γ H H H Γ Γ I I}$$

in which one can note the special symbol for 5 appearing once alone and twice in combination with other symbols.

These numerals were employed for several hundred years prior to the first century B.C. in Attic inscriptions and in public tribute lists and accounts; this explains why the name *Attic* is attached to them. But the name *Herodian*, attached to these numerals, is quite misleading. For Herodian lived in Byzantium around the year A.D. 200, well over five centuries after the appearance of the numerals bearing his name. However, Herodian was a grammarian and he happened to mention the numerals, only once and in a passing way, and thus his name became associated with them.

13° *The origin of the printers' ellipsis.*   In Sanskrit, "zero" was called *sunya*, meaning "empty," and also *sunya-bindu*, meaning "empty-

7

dot," from the fact that in positional notation, and originally on the counting board, a particular position is empty. Such an empty position was shown in written work by a dot. The modern printers' custom of indicating by three dots the omission from a sentence of a word or words goes back to this Indian practice of using a dot to represent zero.

14° *Wise men versus fools.* Thomas Hobbes (1588–1679), the famous English moralist, philosopher, and political scientist, in Part 1 of Chapter 4 of his great work *The Leviathan*, published in 1651 at the conclusion of the civil war that left Cromwell in power, says, "Words are wise men's counters, they do but reckon with them, but they are the money of fools." This is a fine reference to the counters and counting boards of the time.

15° *An unintentional commentary.* In Bailey's *English Dictionary* of 1725 we read, "Ciphers are certain odd Marks and Characters in which Letters are written, that they may not be understood." This may be an unintentional commentary on the general distrust of the Hindu-Arabic numerals in those days.

16° *An old order giving way.* The accounts entered in the *Mayor's Audits* for the City of Bristol of England appear only in Roman numerals until late in the sixteenth century, when Hindu-Arabic numerals begin to make an occasional appearance. In the *Audits* of 1635, the final two pages appear completely in Hindu-Arabic numerals, but then there is a reversion to Roman numerals the following year. From 1640 on, however, Hindu-Arabic numerals only are used.

We witness here an older custom gradually giving way to a newer one, and perhaps a little tug of war between an elderly conservative clerk adhering strictly to traditional methods and a bright young fellow eager to introduce the new mode in figures.

17° *The case for the abacists.* The advocates of the Hindu-Arabic numeral system with its convenient computing algorithms became known as the *algorists*; those defending the Roman numeral system and calculation with it on an abacus became known as the *abacists*.

The battle between the algorists and the abacists was long and some-times bitter. In this battle, the abacists are usually criticized as having been overly conservative and too fearful of change. Of course there was a widespread natural feeling of insecurity with the new numerals; the forms of the numerals were unfamiliar and not standardized, the zero numeral was particularly confusing, and there was much that had to be patiently learned before one could properly use the new system.

But the abacists had a much stronger and much more cogent objection to the new numerals: the new numerals too easily lent them-selves to fraud. One could easily turn a 0 into a 6 or a 9, or turn a 1 into a 4, a 6, a 7, or a 9. Other numeral forms could also be tampered with, and insertion of numerals between or after some already recorded was often quite possible. However, it is difficult to alter a number expressed in Roman numerals, especially if one followed the recom-mended customs of writing the numerals of a number close together, of replacing each final *I* by a *J*, and of running a short horizontal stroke after each final numeral in a number. It was for reasons of such possible falsification that the City Council of Florence, for example, in 1299 issued its ordnance prohibiting under fine the use of the new numerals in financial procedures.

18° *How Xerxes counted his army.*   Herodotus, in Book VII of his *Historia*, tells us how Xerxes once numbered the size of his army on the plain at Doriscus.

> . . . the count of the whole land army showed it to be a million and seven hundred thousand. The numbering was done as follows: a myriad [10,000] men were collected in one place, and when they were packed together as closely as might be, a line was drawn round them; this being drawn, the myriad was sent away, and a wall of stone built on the line reaching up to a man's middle; which done, others were brought into the walled space, till in this way all were counted.

19° *Popular immensities.*   In his article "Large numbers" (*Mathe-matical Gazette*, vol. xxxii, no. 300), J. E. Littlewood mentions a huge stone a cubical mile in size and a million times harder than diamond. Every million years a holy man visits the stone and gives it the very lightest touch. After a period of time, the stone is thus worn away.

The period of time turns out to be something like 10^{35} years, which most people would regard as a vast underestimation.

20° *Overestimation.* Also in his article "Large numbers," Little-wood points out that improbabilities are likely to be overestimated. To illustrate he says, "It is true that I should have been surprised in the past to learn that Professor Hardy had joined the Oxford Group. But one could not say the adverse chance was $1:10^6$. Mathematics is a dangerous profession; an appreciable proportion of us go mad, and then this particular event would be quite likely."

21° *Perfect numbers.* "Perfect numbers like perfect men are very rare," said Decartes. There are only twenty-four perfect numbers known today, and only twelve were known at the time of Descartes.

22° *You can't fool the sergeant.* When the tall inductee presented himself for a physical examination and was asked his height by the examining sergeant the inductee promptly answered:
"I am five feet, seventeen and one-half inches tall, sir."
"You can't kid me, Slim. I know you are more than six feet," snapped the sergeant.—PI MU EPSILON JOURNAL, Nov. 1950, p. 106.

23° *An application of the "least natural number principle."* The untimely passing of Leo Moser on February 9, 1970, at the age of forty-eight, was a real loss to mathematics on the American continent. He was a singularly able and versatile mathematician, and this writer enjoyed years of correspondence with him. As a contrast to his serious work, he was interested in the odd and the amusing in mathematics, and he often sent cute little tidbits to his friends. Here is one, received in a P.S. of a letter dated July 25, 1957.

Here is a new version of an old idea which might amuse you.
Problem: What is the smallest perfect number? *Answer*: 6. Here we have a question which mentions no number explicitly and yet has the unique solution 6. Can one propose for every natural number *n* a problem which mentions no number explicitly and which has *n* as its unique solution?
Solution: Yes! For suppose there are some natural numbers for which

no problem of the required type can be proposed, and consider the problem: "What is the smallest natural number for which there is no problem of the required type?" This yields a contradiction which proves the required result.

24° *Lehmer's table of primes.* In 1914, D. N. Lehmer listed all the primes from 2 to 10,006,721. This was in the days before electronic computers, and the task was a considerable one. The number 10,006,721 is the 664,999th prime number. It is natural to wonder why Lehmer stopped at this particular prime. The explanation is simple. He had defined 1 as a prime also, and thus 10,006,721 was the 665,000th prime number in *his* list. Now his aim was to find all primes not exceeding 10,000,000, and to list them on pages of 5000 primes each. He reached the last prime under 10,000,000 on page 133, and then completed the page, thus listing a total of $(133)(5000) = 665,000$ numbers.

25° *Modern numerals.* Most of us have noticed, and probably wondered about, the peculiar characters (see Figure 2) used for numerals on our bank checks. They are part of a computer data processing practice now widely used in automatic bookkeeping and accounting procedures. The numerals are printed with an ink containing ferromagnetic substances so that they may be read by the magnetic-ink character recognition (MICR) process. The numerals are shaped for better identification by the computer, while at the same time they are readable by us.

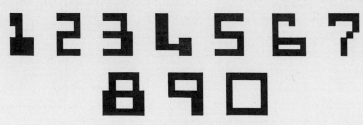

FIGURE 2

The automatic reading method is shown in Figure 3(a). The check is inserted into a reader which slides the check past a permanent magnet, causing the materials in the printed numerals to become magnetically energized. The check then passes over a read head which senses the magnetic density of each numeral and produces corresponding output signals. Matrix decoding is employed to interpret and identify the numerals.

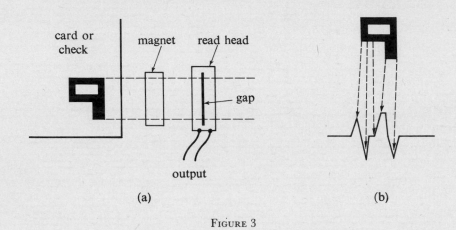

(a)   (b)

FIGURE 3

As shown in Figure 3(b), the signals developed depend upon the width and position of the vertical and horizontal segments of the numeral. For the numeral 9 shown, signals of opposite polarity are produced as the right vertical segment of the 9 approaches and leaves the read head. During the time the horizontal sections pass the head, no magnetic *change* occurs, and the output signals drop to zero level. As the left vertical member of the 9 is reached, a negative and positive voltage change again occurs.

Each numeral produces its own individual signal waveform. The decoding circuits are designed to recognize the particular make-up of a given signal and to establish its numerical value.

26° *Chalk one up for the human being.*   In early November, 1970, subscribers to the various journals handled by Greenwood Publications, Inc., received copies of a letter that commenced as follows:

Dear Subscriber:

I think you should be the first to know that we have fired our computer
—and replaced it with real people who will be able to process and handle
your subscription orders effectively.

Because of the complications and inaccuracies we have experienced
with the automated data processing system, I know that many of you
have not received the total portion of your 1970 subscriptions, and that
is why I have enclosed this postage-paid reply card. I would appreciate
your indicating on the card the issues that you are missing to date so
that we may send them to you. . . .

A great concerted sigh of relief must have escaped on the receipt
of the above letter. For who, these days, has not been harassed and
frustrated by the moronic-minded computer? Heart-rending tales
have been told by subscribers to journals, members of book and record
clubs, mail-order purchasers, people contacting government agencies
in connection with taxes, social security, old-age pensions, medicare,
and so on. The machine would consistently ignore all explanatory
letters and would relentlessly continue to dun and harass the innocent
individuals sucked into its vortex of numbers.

27° *Broadening the college language requirement.* Oakland University in Rochester, Michigan, announced that, beginning with the
1970 winter term, students who want to learn a computer's technical
language can substitute that for the traditional foreign language
requirement.—UPI

28° *A possible hope for the world.* It has often been said that the
pen is mightier than the sword. It might be, then, that a motorized
pen may be mightier than a motorized sword. Herein may lie the
superiority of the electronic computer over the atomic bomb.

29° *A proud boast.* One is hard pressed to think of universal
customs that man has successfully established on earth. There is one,
however, of which he can boast—the universal adoption of the Hindu-
Arabic numerals to record numbers. In this we perhaps have man's
unique worldwide victory of an idea.

MAGIC SQUARES

THERE seems no doubt that magic squares originated in ancient China, whence they found their way to Japan, India, Burma, Siam and other adjacent countries to the south, the Malay Peninsula, and Sumatra. They later played a part in the cabalistic writings of the early Hebrews and they reached the Arabs from India, China, or Persia. Astrologers carried the squares westward, where they influenced medieval European mathematics. There their study prospered fantastically and continued into present times. Today a comprehensive treatise on magic squares and their known remarkable properties would run into many fat volumes.

Magic squares have long been a part of the trappings of fortune tellers. They have been used as charms. They have appeared in fortune bowls, in medicine cups, and on amulets, to protect from illness and plague and to ward off evil. They became linked with medieval alchemy and astrology. Their charm and challenge have attracted a wide array of devotees, initiates to the cult of their study ranging from such eminent mathematicians as Arthur Cayley and Oswald Veblen to such prominent laymen as Benjamin Franklin.

30° *Some terminology and a celebrated unsolved problem.* An *nth order magic square* is a square array of n^2 distinct integers so arranged that the *n* numbers along any row, column, or main diagonal have the same sum, called the *magic constant* of the square. The magic square is said to be *normal* if the n^2 numbers are the first n^2 positive integers; in this case it can be shown that the magic constant is $n(n^2 + 1)/2$. A square is said to be *semimagic* if the numbers along any row or any column have the same sum, but the sum of the numbers along one or each of the main diagonals differs from this sum. A magic square is said to be *symmetric* if any two numbers of the square occupying cells symmetric with respect to the center of the square add to the same sum. In the case of symmetric normal magic squares of order *n*, this sum is always $n^2 + 1$. A magic square is said to be *pandiagonal* (or *nasik*, or *perfect*, or *diabolic*) if, in addition to the numbers along the two main diagonals, the numbers along each broken diagonal also sum to the magic constant of the square. Finally, two magic squares of order *n* are said to be *distinct* if neither one can be obtained from the other by reflections or rotations. From a given magic square, seven others can be obtained by these operations.

Perhaps the most celebrated unsolved problem in connection with normal and with normal pandiagonal magic squares is the determination of the number of them of a given order. In such an enumeration, only distinct squares are counted. This problem has been partially solved. Thus it is known that there is only one normal magic square of order 3, and this is not pandiagonal. There are exactly 880 (distinct) normal magic squares of order 4, and forty-eight of these are pandiagonal. J. Barkley Rosser and Robert J. Walker have shown that there are exactly 3600 normal pandiagonal magic squares of order 5, and that there are no such squares of order n where n is divisible by 2 but not by 4. Thus there is no normal pandiagonal magic square of order 6. This is the extent of our exact knowledge. The number of normal magic squares of order 5 is certainly more than three-quarters of a million, and is estimated to be about fifteen million. It is known that there are more than thirty-eight million normal *pandiagonal* magic squares of order 7, and more than 6.5 trillion of order 8.

The large number of normal magic squares of any given order higher than 3 makes it feasible to construct such a square satisfying some other more or less stringent limitation. The search for such further restricted squares has been intensive.

31° *The lo-shu.* The oldest known example of a magic square is the 3 × 3 square shown in Figure 4. It appears in one of the oldest of the Chinese mathematical classics, the *I-king,* or *Book of Permutations,* a work of uncertain but very ancient origin. But myth claims the square

4	9	2
3	5	7
8	1	6

FIGURE 4

to be even much older, asserting that it was first seen by the Emperor
Yu in about 2200 B.C. as a decoration upon the back of a divine tortoise
along a bank of the Yellow River. Being the earliest, and the simplest,
of all magic squares, this square has appeared more frequently than
any other in connection with fortune telling, charms, amulets, and
cabalistic rites. The square was known in ancient China as the *lo-shu*,
and the numerals in it later took the form of knots in strings, as shown
in Figure 5, in which black knots are used for the even numbers and
white knots for the odd numbers.

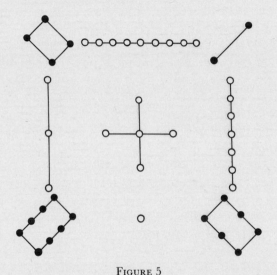

FIGURE 5

32° *The early Hebrews and the lo-shu.* Since the number 15 in
the Hebrew alphabetical numeral system is made up of the first two letters of Jahveh (Jehovah), and since 15 is also the magic constant of the
lo-shu, Eastern Jews of early times found a religious symbol in the
ancient Chinese magic square.

The corner numbers of the lo-shu are even, or female, numbers.
The suppression of these numbers of the square yields a cruciform
arrangement (see Figure 6) of only odd, or masculine, numbers.
This cruciform served as a charm among various eastern peoples.

33° *Talismans and charms.* If the same integer should be added

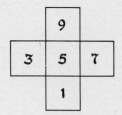

FIGURE 6

to every element of a magic square, the resulting array will, of course, also be magic. Figure 7 shows two magic squares constructed in this way from the lo-shu. The magic constants of these squares are 21 and 18, respectively.

6	11	4
5	7	9
10	3	8

5	10	3
4	6	8
9	2	7

FIGURE 7

Since $2 + 1 = 3$, the first magic square might well serve as a protective symbol on a barn or a ship, for 3 is the number assigned to Jupiter, and Jupiter, as the god of the heavens, manifests himself in atmospheric phenomena. Thus the first magic square might be regarded as a symbol for protection from storm and lightning.

Since $1 + 8 = 9$, the second magic square might serve as a talisman to be worn by a soldier, or as a symbol on a school building, for 9 is the number assigned to Mars, the god of war and the protector of the weak and innocent. In the Maldivian Islands of the Indian Ocean, charms bearing numbers adding to 18 are worn as protective devices by the virgins.

The lo-shu itself might be worn as a talisman for love, or set up in a flower garden, for the magic constant of the lo-shu is 15 and $1 + 5 = 6$, the number of Venus, goddess of love and of the garden.

17

34° *The earliest fourth-order magic square.*   The earliest recorded magic square of the fourth order is found in an eleventh or twelfth century Jaina inscription at Khajuraho, in India, and is shown in Figure 8. This square exhibits a somewhat advanced knowledge of magic squares, for it possesses further "magic" properties. First of all, it is pandiagonal—for example, the numbers, 2, 12, 5, and 15,

7	12	1	14
2	13	8	11
16	3	10	5
9	6	15	4

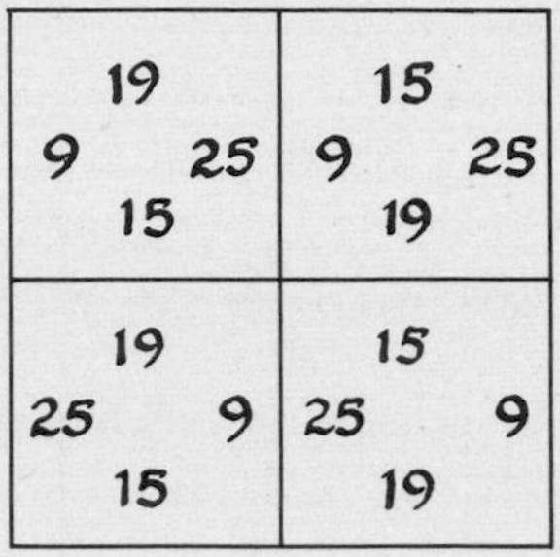

19		15	
9	25	9	25
15		19	
19		15	
25	9	25	9
15		19	

FIGURE 8

and the numbers 2, 3, 15, and 14, are along broken diagonals, and these sum to the magic constant 34. Further, the four subsquares formed by drawing the horizontal and vertical midlines of the given square are interestingly related to one another, as can be seen from Figure 8, where in the top left subsquare the 19 is 7 + 12, the 15 is 2 + 13, the 9 is 7 + 2, and the 25 is 12 + 13. This square is perhaps the earliest instance showing some of the further fantastic elaborations and limitations that can be assigned to a normal magic square.

35° *Albrecht Dürer's magic square.*   Of the artist-mathematicians of the sixteenth century, Albrecht Dürer stands out prominently. He was born at Nürnberg in 1471 and died in the same city in 1528. Famous as a painter and an engraver, he also showed considerable interest and ability in various areas of mathematics, as is shown by his treatises on geometry, fortifications, and human proportions. His geometry was the first printed work on higher plane curves. His approximate solution of the trisection problem and of the construction of a regular nonagon and a regular heptagon are of interest to high school geometry students.

Dürer's interest in mathematics is also evident from some of his engravings, in particular his famous etching *Melencolia*. In this engraving appear various mathematical items, such as a sphere, a compass and straightedge, a proportional compass, a polyhedral solid, and a magic square of the fourth order. The magic square, which was one of the first magic squares to be printed, is pictured in Figure 9. In addition to the usual "magic" properties of such squares, it has the following curious properties:

16	3	2	13
5	10	11	8
9	6	7	12
4	15	14	1

Figure 9

1. The date, 1514, in which the *Melencolia* engraving was made, appears in the two middle cells of the bottom row.
2. The sum of the squares of the numbers in the first two rows (columns) is equal to the sum of the squares of the numbers in the last two rows (columns).
3. The sum of the squares of the numbers in the first and third rows (columns) is equal to the sum of the squares of the numbers in the second and fourth rows (columns).
4. The sum of the numbers in the diagonals is equal to the sum of the numbers not in the diagonals.
5. The sum of the squares of the numbers in the diagonals is equal to the sum of the squares of the numbers not in the diagonals.
6. The sum of the cubes of the numbers in the diagonals is equal to the sum of the cubes of the numbers not in the diagonals.

36° *A pandiagonal magic square on a torus.* A dramatic way to exhibit the properties of a pandiagonal magic square is to draw the square on a torus. We can accomplish this by taking the pandiagonal magic square, as in Figure 10(a), and first forming a cylinder by bringing together the top and the bottom of the square as in Figure 10(b), and then finally forming a torus by stretching and bending the cylinder as in Figure 10(c). All rows, columns, and diagonals (both main and broken) of the square become closed loops on the torus.

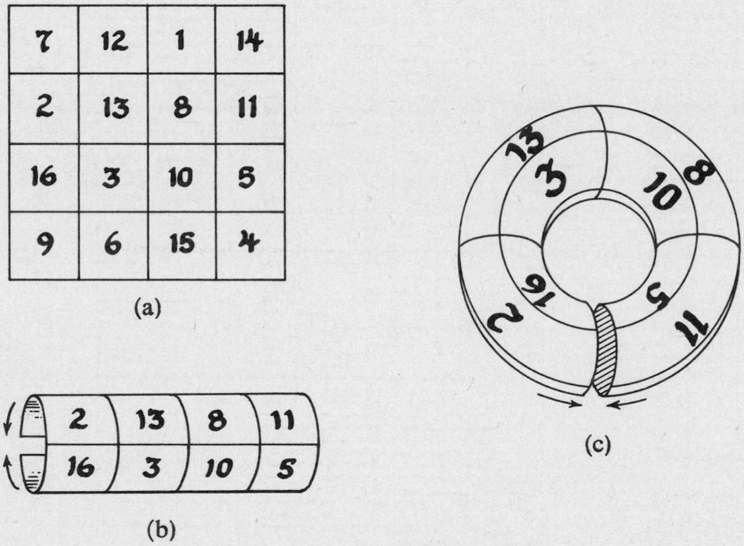
7 12 1 14
2 13 8 11
16 3 10 5
9 6 15 4
(a)
2 13 8 11
16 3 10 5
(b)
13 3 10 8
16 2 5 11
(c)

FIGURE 10

If, for a fourth-order normal pandiagonal magic square rolled and bent into a torus in the above fashion, we start with any number and move two numbers away in any direction along a diagonal, we arrive at the same number. Two numbers related in this way are called *antipodes*, and the sum of any two antipodes is 17. The numbers along any loop (diagonal or orthogonal) sum to 34; any square group of four adjacent numbers sums to 34.

37° *A pandiagonal magic square on a plane.* If we form a mosaic on a plane by fitting together a large number of duplicate pandiagonal squares of order n (see Figure 11), we obtain a field of squares on which any $n \times n$ group of numbers is pandiagonal. Any n adjacent numbers of the field, taken vertically, horizontally, or diagonally, yield the same magic sum.

13	4	9	6	15	4	9	6	15	4	9	6	15	4	9	6	15
1	14	7	12	1	14	7	12	1	14	7	12	1	14	7	12	1
8	11	2	13	8	11	2	13	8	11	2	13	8	11	2	13	8
10	5	16	3	10	5	16	3	10	5	16	3	10	5	16	3	10
15	4	9	6	15	4	9	6	15	4	9	6	15	4	9	6	15
1	14	7	12	1	14	7	12	1	14	7	12	1	14	7	12	1
8	11	2	13	8	11	2	13	8	11	2	13	8	11	2	13	8
10	5	16	3	10	5	16	3	10	5	16	3	10	5	16	3	10
15	4	9	6	15	4	9	6	15	4	9	6	15	4	9	6	15
1	14	7	12	1	14	7	12	1	14	7	12	1	14	7	12	1
8	11	2	13	8	11	2	13	8	11	2	13	8	11	2	13	8
10	5	16	3	10	5	16	3	10	5	16	3	10	5	16	3	10
15	4	9	6	15	4	9	6	15	4	9	6	15	4	9	6	15
1	14	7	12	1	14	7	12	1	14	7	12	1	14	7	12	1

FIGURE 11

38° *A pandiagonal magic square on a hypercube.* We can map a fourth-order pandiagonal magic square onto a hypercube by transferring the sixteen numbers of the square to the sixteen vertices of the hypercube, as shown in Figure 12, where we employ the familiar two-dimensional projection of a hypercube. The sum of the numbers at the vertices of each of the twenty-four square faces of the hypercube is 34. Antipodal points of the magic square appear as diagonally opposite vertices of the hypercube.

؟	12	1	14
2	13	8	11
16	3	10	5
9	6	15	4

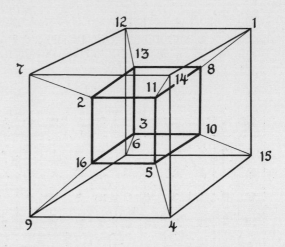

FIGURE 12

39° *A magic square in architectural decoration.* One of the most elaborate instances of a magic square employed in architectural decoration is one of the ninth order chiseled into a wall in the Villa Albani in Rome.

40° *Magic squares of odd order.* Simon de la Loubère,* when envoy of Louis XIV to Siam in 1687–1688, learned a simple method for finding a magic square of any odd order. Let us illustrate the method by constructing one of the fifth order. Draw a square and divide it into 25 cells (see Figure 13). Border the square with cells along the top and right edges, and shade the added cell in the top right corner. Write 1 in the middle top cell of the original square. The general rule is then to proceed diagonally upward to the right with the successive integers. Exceptions to this general rule occur when such an operation takes us out of the original square or leads us into a cell already occupied. In the former situation we get back into the original square by shifting

---

* Not to be confused with Antoine de la Loubère (1600–1664), a Jesuit lecturer on mathematics, rhetoric, and religion, who showed interest in the study of both planar and spatial curves. Simon de la Loubère later in life wrote a work on the solution of equations.

clear across the square, either from top to bottom or from right to left, as the case may be, and then continue with the general rule. In the second situation we write the number in the cell immediately beneath the one last filled, and then continue with the general rule. The shaded cell is to be regarded as occupied.

	18	25	2	9	
17	24	1	8	15	17
23	5	7	14	16	23
4	6	13	20	22	4
10	12	19	21	3	10
11	18	25	2	9	

FIGURE 13

Thus, in our illustration, the general rule would place 2 diagonally upward from 1 in the fourth cell bordered along the top edge. We must, therefore, shift the 2 to the fourth cell in the bottom row of the original square. When we come to 4, it falls in the third cell up bordered along the right edge. It must, therefore, be written clear across to the left in the third cell up in the first column of the original square. The general rule would place 6 in the cell already occupied by 1. It is accordingly written in the cell just below that occupied by the last written number, 5. And so on.

The reader may now care to construct, by de la Loubère's rule, a magic square of order seven. The lo-shu is essentially the 3 × 3 magic square as constructed by de la Loubère's method.

41° *Magic squares of doubly-even order.* There is an easy way to construct magic squares of doubly-even order, that is, magic squares whose orders are a multiple of 4. Consider, first of all, a square of order

23

4 and visualize the diagonals as drawn (see Figure 14). Beginning in the upper left corner, count across the rows from left to right in descending succession, recording only the numbers in cells not cut by a diagonal.

	2	3	
5			8
9			12
	14	15	

16	2	3	13
5	11	10	8
9	7	6	12
4	14	15	1

FIGURE 14

Now, beginning at the lower right corner, count across the rows from right to left in ascending succession, recording only the numbers in the cells that are cut by a diagonal. The resulting magic square is little different from Dürer's square of Item 35°.

The same rule applies to any magic square of order $4n$ if we visualize, as drawn in, the diagonals of all the n^2 principal 4×4 sub-blocks. Figure 15 shows the construction of an 8×8 magic square by this rule. The reader may now care to construct a magic square of order 12.

	2	3			6	7	
9			12	13			16
17			20	21			24
	26	27			30	31	
	34	35			38	39	
41			44	45			48
49			52	53			56
	58	59			62	63	

64	2	3	61	60	6	7	57
9	55	54	12	13	51	50	16
17	47	46	20	21	43	42	24
40	26	27	37	36	30	31	33
32	34	35	29	28	38	39	25
41	23	22	44	45	19	18	48
49	15	14	52	53	11	10	56
8	58	59	5	4	62	63	1

FIGURE 15

42° *Doubly-magic and trebly-magic squares.* For certain values of *n* it is possible to construct an *n*th order magic square such that if each number of the array is squared, the resulting array will also be magic. Figure 16 shows an example, due to M. H. Schots, of a normal doubly-magic pandiagonal square of order 8. The magic constant of the square is 260; the magic constant of the square obtained by squaring each element is 11180.

16	41	36	5	27	62	55	18
26	63	54	19	13	44	33	8
1	40	45	12	22	51	58	31
23	50	59	30	4	37	48	9
38	3	10	47	49	24	29	60
52	21	32	57	39	2	11	46
43	14	7	34	64	25	20	53
61	28	17	56	42	15	6	35

FIGURE 16

Figure 17 shows an example, due to R. V. Heath, of a normal doubly-magic square of order 9. There are no normal doubly-magic

70	75	59	15	26	1	38	49	36
11	22	9	43	48	32	69	80	58
42	53	28	65	76	63	16	21	5
57	68	79	8	10	24	31	45	47
4	18	20	30	41	52	62	64	78
35	37	51	58	72	74	3	14	25
77	61	66	19	6	17	54	29	40
27	2	13	50	34	39	73	60	71
46	33	44	81	56	67	23	7	12

FIGURE 17

25

squares of order less than 8.

A square which is magic for the original numbers, for their squares, and for their cubes is said to be trebly magic. There exists normal trebly-magic squares of orders 64, 81, and 128.

43° *Prime magic squares.* A magic square is said to be *prime* if each of its numbers is a prime, 1 being considered as a prime. The famous British puzzle expert, H. E. Dudeney, gave the prime square of Figure 18(a). E. Bergholt and C. D. Shuldham gave the prime square of Figure 18(b). Here occur the first 13 odd primes, along with the primes 47, 53, and 71. Prime magic squares of orders 5, 6, · · ·, 12 have been constructed by H. A. Sayles and J. N. Muncey. Muncey's prime magic square of order 12 is particularly interesting in that its 144 numbers are the first 144 odd primes: 1, 3, 5, 7, 11, · · ·, 827. It has been shown that the first n^2 odd primes cannot form a magic square when $n < 12$.

67	1	43
13	37	61
31	73	7

(a)

3	71	5	23
53	11	37	1
17	13	41	31
29	7	19	47

(b)

FIGURE 18

Figures 19(a) and 19(b) show prime magic squares of the third and fourth orders wherein the number 1 is not admitted as a prime number. In the October 1961 issue of the *Recreational Mathematics Magazine* (p. 28) appears a prime magic square of this sort of order 13. It has been conjectured that there exist infinitely many prime magic squares of this sort for every order $n > 8$.

44° *Fibonacci magic squares.* The sequence of numbers 1, 1, 2, 3, 5, 8, · · ·, wherein, after the first two numbers are written, each number

569	59	449
239	359	479
269	659	149

17	317	397	67
307	157	107	227
127	277	257	137
347	47	37	367

(a)   (b)

FIGURE 19

is the sum of the two immediately preceding numbers, is called the *Fibonacci sequence*. It has been proved that there are no magic squares composed only of numbers of the Fibonacci sequence. There are, however, magic squares composed only of such numbers and sums of pairs of such numbers. (For some of the esoteric properties of the Fibonacci sequence, see Item 114° of *In Mathematical Circles*.)

45° *Magic cubes.* An *nth order magic cube* is a cubical array of n^3 distinct integers so arranged that the n numbers along any row, column, file, or main diagonal have the same sum, called the *magic constant* of the cube. The magic cube is said to be *normal* if the n^3 numbers are the first n^3 positive integers. In this case the magic constant is $n(n^3 + 1)/2$, and it occurs in $3n^2 + 4$ ways. Normal magic cubes of any odd or any doubly-even order can be constructed by extensions of the methods used for normal magic squares. A magic cube is said to be *pandiagonal* if not only its four main diagonals, but all broken diagonals as well, sum to the magic constant. In a pandiagonal magic cube the magic constant occurs $7n^2$ times.

One of the most remarkable examples of a normal pandiagonal magic cube was given by R. V. Heath and is illustrated in Figure 20. Here an eighth-order normal magic square is cut into quarters, each quarter serving as a layer of a fourth-order normal pandiagonal magic cube. Each layer is itself a fourth-order magic square. The original eighth-order magic square has the additional property that if either set of alternate rows and either set of alternate columns be deleted—

27

and this can be done in four ways—the remaining 16 numbers form a fourth-order magic square of magic constant 130.

(first layer)   (second layer)

1	8	61	80	48	41	20	21
62	59	2	7	19	22	47	42
52	53	16	9	29	28	33	40
15	10	51	54	34	39	30	27
32	25	36	37	49	56	13	12
35	38	31	26	14	11	50	55
45	44	17	24	4	5	64	57
18	23	46	43	63	58	3	6

(fourth layer)   (third layer)

FIGURE 20

46° *Brains versus machines.* An *nth order Latin square* is an $n \times n$ array such that each of n distinct symbols appears once and only once in each row and each column of the array. An *nth order Eulerian square* is an $n \times n$ array of n^2 distinct elements wherein the (i, j)th element (that is, the element in the ith row and the jth column) is an ordered pair of symbols, the first one being the (i, j)th element of one Latin square and the second one being the (i, j)th element of a second Latin square. For example, in Figure 21 there are two fifth-order Latin squares in each of which 0, 1, 2, 3, 4 are the five distinct symbols. From these two Latin squares we obtain the fifth-order Eulerian square (since all 25 elements in this square are distinct) shown in Figure 22.

For a couple of hundred years methods have been known for constructing Eulerian squares of odd order and of doubly-even order

```
3   1   4   2   0        4   2   0   3   1
2   0   3   1   4        0   3   1   4   2
1   4   2   0   3        1   4   2   0   3
0   3   1   4   2        2   0   3   1   4
4   2   0   3   1        3   1   4   2   0
```

FIGURE 21

```
34  12  40  23  01
20  03  31  14  42
11  44  22  00  33
02  30  13  41  24
43  21  04  32  10
```

FIGURE 22

(that is, of order $n = 4m$). After much futile effort to construct an Eulerian square of singly-even order (that is, of order $n = 4m + 2$), Euler conjectured in 1782 that none exists. This conjecture is trivial for $n = 2$, and in 1900 G. Tarry verified it for $n = 6$ by the exhausting method of systematically examining all possibilities. For the next case, $n = 10$, the number of possibilities is so enormous that Tarry's procedure is utterly worthless. Indeed, even an electronic computer proved to be of no use. One of these machines once spent 100 hours searching for a tenth-order Eulerian square, failed to find any, and still had not even begun to scratch the surface in its examination of possibilities. It was not until 1959 that further progress was made; in that year R. C. Bose, S. S. Shrikhande, and E. T. Parker disproved Euler's conjecture. Parker gave the tenth-order Eulerian square shown in Figure 23. A year later it was shown that Euler's conjecture is wrong for all $n > 6$; that is, there exist nth-order Eulerian squares for all orders n except $n = 2$ and $n = 6$.

29

00	47	18	76	29	93	85	34	61	52
86	11	57	28	70	39	94	45	02	63
95	80	22	67	38	71	49	56	13	04
59	96	81	33	07	48	72	60	24	15
73	69	90	82	44	17	58	01	35	26
68	74	09	91	83	55	27	12	46	30
37	08	75	19	92	84	66	23	50	41
14	25	36	40	51	62	03	77	88	99
21	32	43	54	65	06	10	89	97	78
42	53	64	05	16	20	31	98	79	87

FIGURE 23

The fifth-order and tenth-order Eulerian squares exhibited in Figures 22 and 23 are semimagic squares. Every Eulerian square leads to at least a semimagic square.

47° *Origin of Eulerian squares.* Eulerian squares originated in a problem that Euler stated as follows: "A very curious question is the following: A group of 36 officers of 6 different ranks and from 6 different regiments are to be arranged in a square in such a manner that each row and each column contains 6 officers from different regiments and of different ranks." Euler was unable to obtain a solution to the problem.

Since the desired square array of officers would be an Eulerian square of order 6, we know (see Item 46°) that the problem has no solution.

30

MATHEMATICALLY MOTIVATED DESIGNS

A WORTHY, enjoyable, and not at all difficult project would be to assemble a booklet of unusual and attractive designs originating in certain mathematical problems and puzzles. Following are some illustrations of the idea. Perhaps some reader may feel motivated to continue the project; the material would be of value in high school mathematics laboratories.

48° *Some artistic applications of the lo-shu.*   Not many years ago, before the advent of jet plane travel, the 3×3 magic square known as the lo-shu (see Figure 4, Item 31°) was commonly seen on the play decks of large transoceanic passenger ships, for scoring in such games as shuffleboard.

One person has used the lo-shu in its knotted-string representation (see Figure 5, Item 31°) as a pleasing decoration on his bedroom ceiling, motivated by the connection of the lo-shu with the number 6 assigned to Venus, the goddess of love (see Item 33°).

Another person, who designed a beach house with oriental motifs, employed the knotted-string representation of the lo-shu to light the central room of the house. The knots appear as tiny lights in small cylindrical recesses in the ceiling, blue lights for the odd (or masculine) numbers and pink lights for the even (or feminine) numbers.

49° *Magic traces.*   The American architect and occultist, Claude Fayette Bragdon (who died in 1946), became fascinated by the observation that the line segments traced from cell to cell, in the sequential order of the numbers, on many magic squares produce a pleasing artistic pattern. Sometimes it is advisable to trace the line segments sequentially connecting only the odd (masculine) or only the even (feminine) numbers. Mr. Bragdon used designs obtained in this way for book covers, textile patterns, and architectural ornaments. He decorated the chapter headings of his autobiography, *More Lives Than One*, with these designs, and he constructed the ventilating grill in the ceiling of the Chamber of Commerce of his home city of Rochester, New York, as the trace obtained from the lo-shu.

Figure 24 shows the design yielded by the fifth-order magic square constructed by de la Loubère's method. This pattern, drawn on a screen or on a scarf, or utilized as grille work for a gate, would be quite attractive, and few beholders could guess its origin. The author was once presented with a gold tie clasp bearing this interesting design engraved on it. Figures 25(a) and 25(b) show the "masculine" and the "feminine" traces for the same magic square.

Figure 26 shows the "masculine" and the "feminine" traces, superposed, of the lo-shu; it could serve as a symbol of marriage and could fittingly decorate the door, and the headboard of the bed, of a honeymoon suite.

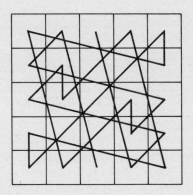

FIGURE 24

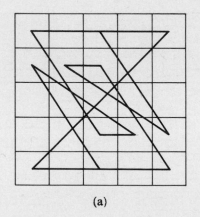

(a)

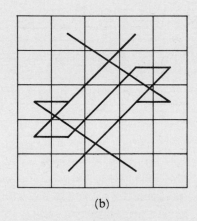

(b)

FIGURE 25

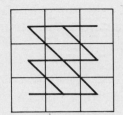

FIGURE 26

50° *Re-entrant knight's path.* One of the challenging problems on an ordinary chessboard is that of determining a path a knight might take so as to land on each square of the board once and only once, returning to the starting square on the last jump. One of the prettiest solutions to this problem was given by P. M. Roget in 1840 and is shown in Figure 27. The path traced by the knight yields a singularly attractive design for a screen, a tapestry, or iron grille work.

34	51	32	15	38	53	18	3
31	14	35	52	17	2	39	54
50	33	16	29	56	37	4	19
13	30	49	36	1	20	55	40
48	63	28	9	44	57	22	5
27	12	45	64	21	8	41	58
62	47	10	25	60	43	6	23
11	26	61	46	7	24	59	42

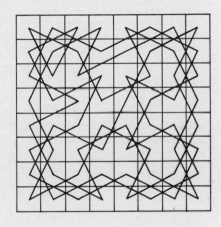

FIGURE 27

The total possible number of distinct re-entrant knight's paths on a chessboard is not known, but it is a very large number lying somewhere between the number of combinations of 168 things taken 63 at a time as an upper bound and 122,802,512 as a lower bound. Since the problem has so many solutions and since there are so many magic

33

squares of order 8, it is natural to wonder if there is a re-entrant knight's path on the ordinary chessboard that also yields a magic square of order 8. Figure 28 shows a re-entrant knight's path that yields a semi-magic square of order 8—that is, the rows and the columns add to the same sum, but the two main diagonals do not. This latter design would make an attractive pattern on a bedspread. It would also be attractive laid out on the floor of a patio.

63	22	15	40	1	42	59	18
14	39	64	21	60	17	2	43
37	62	23	16	41	4	19	58
24	13	38	61	20	57	44	3
11	36	25	52	29	46	5	56
26	51	12	33	8	55	30	45
35	10	49	28	53	32	47	6
50	27	34	9	48	7	54	31

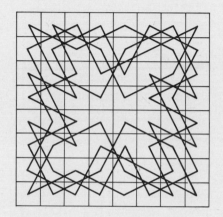

FIGURE 28

35	40	63	54	15	12	17	26
62	53	34	39	18	27	14	11
41	36	55	64	13	16	25	28
52	61	38	33	24	19	10	5
37	42	51	56	1	6	29	20
60	57	48	45	32	23	4	9
43	46	59	50	7	2	21	30
58	49	44	47	22	31	8	3

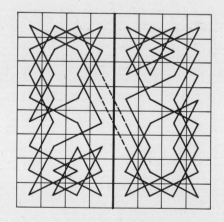

FIGURE 29

34

35	38	55	58	11	14	31	18
54	57	34	37	32	17	10	13
39	36	59	56	15	12	19	30
60	53	40	33	20	29	16	9
41	48	61	52	1	8	21	28
62	51	44	47	24	27	4	7
45	42	49	64	5	2	25	22
50	63	46	43	26	23	6	3

FIGURE 30

51° *Half-board solutions.* The two solutions of the re-entrant knight's path illustrated in Figures 29 and 30 are remarkable in that the right half of the board is covered first, then the left half of the board. The first of these solutions is due to the great Swiss mathematician Leonhard Euler; the second solution is due to P. M. Roget. Each solution is especially well suited for the decoration of a two-part folding screen. They would also be attractive on bedspreads for double beds.

15	20	17	36	13	64	61	34
18	37	14	21	60	35	12	63
25	16	19	44	5	62	33	56
38	45	26	59	22	55	4	11
27	24	39	6	43	10	57	54
40	49	46	23	58	3	32	9
47	28	51	42	7	30	53	2
50	41	48	29	52	1	8	31

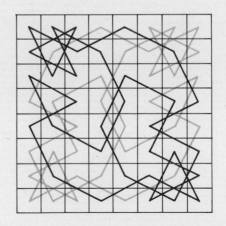

FIGURE 31

35

52° *Magic re-entrant half-board knight's paths.*   Each of the two re-entrant knight's paths shown in Figure 31 covers exactly 32 cells of the board. They are remarkable in that together they constitute a magic square. The two paths, done in two different colors, would furnish a very attractive pattern for a blanket or a bedspread.

53° *Re-entrant strong-knight paths.*   Whereas an ordinary knight on the chessboard moves one square diagonally followed by one square nondiagonally, a *strong knight* moves one square diagonally followed by two squares nondiagonally. It follows that if a strong knight starts on, say, a black square of a chessboard, it is ever after restricted to black squares. This raises the question of whether we can cover a chessboard by two re-entrant strong-knight's paths, one knight covering the black squares of the board and the other covering the red squares. Figure 32 shows a beautiful solution to the problem, which again suggests a very attractive design for a blanket or a bedspread. (*Note*: A strong knight is an instance of a *leaper* in fairy chess.)

FIGURE 32

36

54° *Magic re-entrant king's path.* Figure 33 shows a re-entrant king's path on a chessboard, which also yields a magic square. It is a splendid design for a bed blanket, and it looks like an authentic American Indian pattern.

61	62	63	64	1	2	3	4
60	11	58	57	8	7	54	5
12	59	10	9	56	55	6	53
13	14	15	16	49	50	51	52
20	19	18	17	48	47	46	45
21	38	23	24	41	42	27	44
37	22	39	40	28	26	43	28
36	35	34	33	32	31	30	29

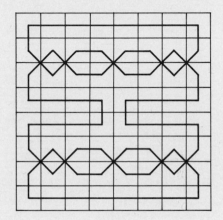

FIGURE 33

55° *A conversation piece.* Henry Ernest Dudeney (1857–1931), by all odds England's foremost inventor of puzzles, was spectacularly adept in the art of geometrical dissections. His best-known achievement in this field is his four-piece dissection of an equilateral triangle into a square. A five-piece solution of this problem had been known for some time, and it was generally believed that this could not be bettered. Dudeney's discovery was made about 1902, and has since become famous in dissection literature. Figure 34 shows the dissection. The segments *AD, DB, BE, EC, FG* are all equal to half the side of the triangle; *EF* is equal to the side of the equivalent square; *DJ* and *GK* are each perpendicular to *EF*.

If the four pieces 1, 2, 3, 4 are successively hinged to one another at the points *D, E, G*, then, holding piece 1 fixed and swinging the connected set of pieces 4–3–2 counterclockwise (see Figure 35), the equilateral triangle is neatly carried into the square. A set of four connected tables has been built upon this fact; swinging the tables in one direction causes the tops to fit together into a single equilateral

37

triangular table, and swinging them in the other direction causes the tops to fit together into a single square table. The table is adaptable for card games requiring either three or four players. It makes an interesting piece of furniture, and, if nothing else, is a fine conversation piece.

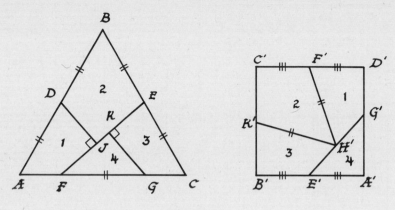

FIGURE 34

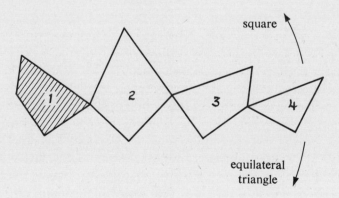

FIGURE 35

56° *A patio mosaic.* L. V. Lyons has used Dudeney's construction given in the preceding Item to obtain a very attractive dissection of the plane into a mosaic of interlocking equilateral triangles and squares, as pictured in Figure 36. This artistic pattern is excellent for a patio floor, or for use somewhere in a university mathematics building.

38

FIGURE 36

57° *The octagon puzzle.* Dudeney bettered several long-established records in polygonal dissection. He was the first, for example, to dissect a regular pentagon into a square with only six pieces, and to dissect a given square into three equal squares with only six pieces. Very curious along this line is a statement made by Alice Dudeney (Mrs. H. E. Dudeney) in the preface to *Puzzles and Curious Problems*, by H. E. Dudeney, published in 1932, after the author had died. The book was reprinted in 1936, 1941, and 1948, revised by James Travers. The statement is: "It is remarkable that the regular octagon can be cut into as few as four pieces to form a corresponding square." The implication is that such a four-piece dissection was another of Dudeney's triumphs. But no one has since been able to find such a solution. A beautiful five-piece dissection, illustrated in Figure 37, was published by Travers in 1933. It is strange that Travers, who revised the book in question, made no correction or comment on Alice Dudeney's unverifiable statement. It is difficult to believe that a four-piece solution of the problem is possible.

39

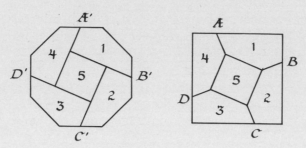

FIGURE 37

We have, in Travers' dissection, the possibility of another unusual set of tables. Four identical tables (1, 2, 3, 4 in Figure 37) can be placed tightly about a square pillar to form a single surrounding octagonal table or a single surrounding square table.

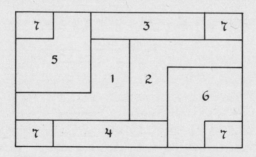

FIGURE 38

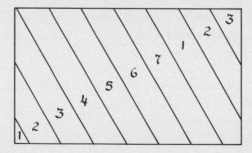

FIGURE 39

40

58° *A toral tulip garden.*   P. J. Heawood was the first to prove that seven is the least number of colors needed to color all maps on a torus. One establishes this theorem by simply exhibiting a map on a torus that requires seven colors, since it can be shown that seven colors are sufficient. Many simple maps of this sort can be devised; Figures 38 and 39 show two such maps. In each of these figures one is first to imagine the rectangle rolled into a tube by joining together the top and bottom edges of the rectangle, and then to imagine the tube bent into a torus by joining together the two circular ends of the tube. Each map is then seen to contain seven countries so arranged that each country shares a boundary with each of the other six, thus necessitating seven colors to color the map properly.

Each of Figures 38 and 39 can be cleverly utilized for a rectangular flower bed, say a bed of tulips, wherein tulips of seven different colors are planted to form the seven different countries of the map.

The figures can also serve as designs for a courtyard or entrance hall in a mathematics building of a university. Here the seven different countries could be indicated by seven different colors of cement or floor covering.

59° *A flower garden on a projective plane.*   Heinrich Tietze proved that six is the least number of colors needed to color all maps on a projective plane. Since a projective plane can be represented by a circular disk in which diametrically opposite points on the rim of the disk are identified with one another, any one of Figures 40(a), 40(b), 40(c) represents a six-color map on a projective plane (each of the six countries shares a boundary with each of the other five). As in the preceeding Item, any one of these three maps can serve admirably for the design of a flower garden—this time circular and requiring six different colors of flowers.

60° *Schlegel flower garden.*   While on the subject of designs for flower gardens, we might mention some so-called Schlegel diagrams (named after Victor Schlegel, 1843–1905) of the five regular polyhedra. A Schlegel diagram of such a polyhedron (or, more generally, of any convex polyhedron) can be obtained by a central projection of the polyhedron from an exterior point sufficiently close to the centroid of a selected face of the polyhedron. The projection of the selected face contains the projections of all the other faces. Figure 41 shows such Schlegel diagrams of the five regular polyhedra. Each of these diagrams could serve nicely as the design of a flower bed.

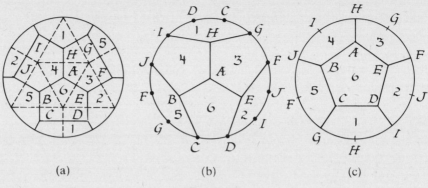

(a)   (b)   (c)

FIGURE 40

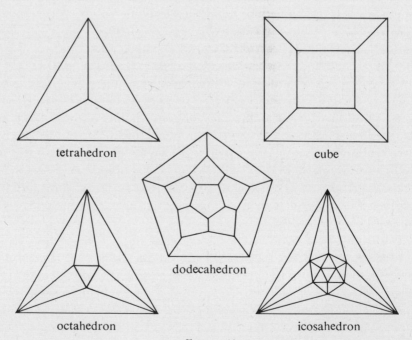

tetrahedron   cube

dodecahedron

octahedron   icosahedron

FIGURE 41

42

61° *Professor Tucker's flag.*   Figure 42 shows another toral map (see Item 58°), here represented on a square. We certainly have in this map a very interesting and attractive design for a lady's head scarf.

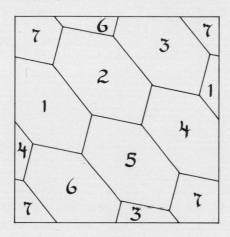

FIGURE 42

By a horizontal stretch, one can convert the square map of Figure 42 into a rectangular map. About this rectangular map, Professor Albert Tucker of Princeton University once remarked, many years ago during a course in elementary combinatorial topology that he was teaching, "If I were a dictator, I would fashion my flag after this map."

62° *More patio designs.*   In 1925, Z. Morón noted (see Figure 43) that a 32 × 33 rectangle can be dissected into nine squares, no two of which are equal. This raised the question of whether a square can be dissected into a finite number of squares, no two of which are equal. This latter problem was felt to be impossible, but such turned out not to be the case. The first published example of a square dissected into unequal squares appeared in 1939; the dissection was given by R. Sprague of Berlin, and contained fifty-five subsquares. In 1940, R. C. Brooks, C. A. B. Smith, A. H. Stone, and W. T. Tutte, in a joint paper, published a dissection containing only 26 pieces. These men ingeniously established a connection between the problem of dissec-

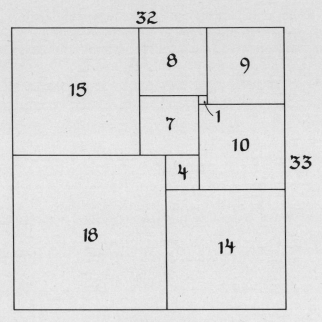

FIGURE 43

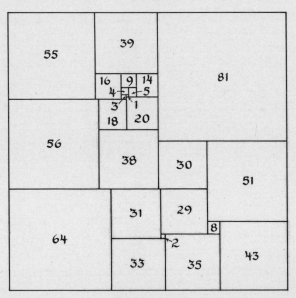

FIGURE 44

tion and certain properties of currents in electrical networks. In 1948, T. H. Willcocks published the 24-piece dissection of a square into unequal squares pictured in Figure 44, and this dissection is today the record so far as least number of pieces is concerned. Surely, somewhere in the planning and designing of a mathematics building at a university, one or other of these two squarings should appear.

63° *In the foyer of the mathematics building.* It would seem fitting, when considering the construction of a new mathematics building, to incorporate into the building a number of attractive designs and motifs representing the different areas of mathematical study. Some possible suggestions have already been made in a few of the preceding Items.

Now in the area of projective geometry there are few prettier or more fertile theorems than the famous *Pascal mystic hexagram theorem,* which states that "the three points of intersection of the three pairs of opposite sides of a hexagon inscribed in a conic section are collinear." Why not, then, in the entrance hall of the new building, have this theorem attractively represented as in Figure 45, using, perhaps, inlaid brass bands for the lines of the figure?

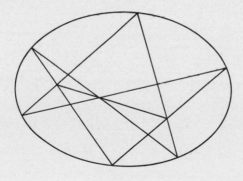

FIGURE 45

64° *Room numbers.* In the new mathematics building referred to in Item 63°, one might employ some attractive numeral system, in addition to the usual Hindu-Arabic system, for numbering the doors of the rooms and offices of the building. The choice of the additional numeral system might depend upon the country or place of the university. In Mexico it would be natural to choose the ancient Mayan numeral system, and in Egypt the ancient Egyptian hieroglyphic system. Other possibilities are Babylonian cuneiform, Hebrew, Chinese (traditional or scientific), Greek (alphabetic or Herodianic), Persian, and Roman. It should be particularly fitting in an institution of learning to have such little evidences of the blending of an old and a new culture.

65° *Tying gift boxes.* Frequently gift boxes are artfully tied with a ribbon as indicated in Figure 46; two parallel and more-or-less diagonally directed segments of the ribbon can be seen on both the top and the bottom of the box. If the length, width, and depth of the box

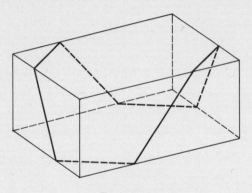

FIGURE 46

are represented by *a*, *b*, and *c*, respectively, it can be shown that the length of the ribbon (not counting any knot or bow) is

$$d = 2\sqrt{(a + c)^2 + (b + c)^2},$$

and that the ribbon can be shifted into parallel positions without stretching. (The tangent of the angle at which the ribbon intersects the edges of the box is $(a + c)/(b + c)$ or $(b + c)/(a + c)$.) Many large department stores use these facts to secure boxes, quickly and efficiently, with an endless loop of slightly elastic tape, of which the length of the unstretched loop is a trifle less than *d*.

66° *A Pythagorean theorem design.* Sometimes the diagram of a geometrical proof can be utilized as a design. For example, there is a simple "dissection" proof of the Pythagorean theorem that runs as follows. The square on the hypotenuse of the right triangle is dissected, as shown in Figure 47(a), into four triangles congruent to the given right triangle and a square of side equal to the difference of the legs

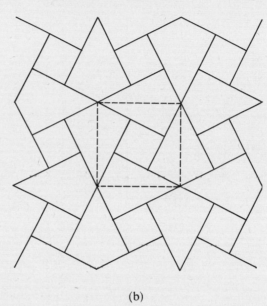

(a)

(b)

FIGURE 47

of the right triangle. Denoting the legs and the hypotenuse of the right triangle by a, b, c, where $a > b$, we then have

$$c^2 = 4(ab/2) + (a - b)^2 = 2ab + (a^2 + b^2 - 2ab) = a^2 + b^2,$$

and the Pythagorean theorem is established. Figure 47(b) shows the diagram of this proof employed as the basis of an attractive lattice design. This lattice was actually used by the Arabs in the Middle Ages.

67° *A maple-leaf tile.* One usually sees a floor tiled with square, rectangular, equilateral triangular, or regular hexagonal tiles. Tiling a floor with tiles of some other single shape is interesting and a bit

47

challenging. Figure 48(a) shows a simple tile, roughly resembling a maple leaf, that can be used for tiling a floor (see Figure 48(b)). Such a floor might be particularly attractive in a public building in Canada.

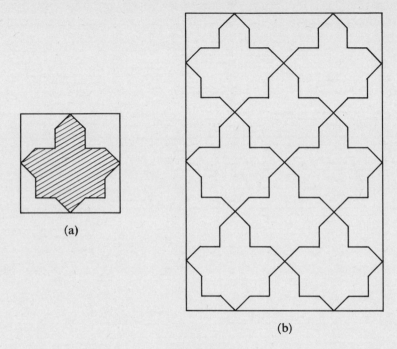

(a)

(b)

FIGURE 48

68° *Another maple-leaf tile.* Figure 49 shows a somewhat more complicated maple-leaf tile and a more complicated tiling with it. This tiling appears in the Alhambra, and a tile of this shape has been manufactured for the present-day market.

69° *Some other tiles.* The six parts of Figure 50 show a number of other attractive tiles. Some of these designs can also serve for iron grille work. The tiling shown in Figure 50(b) has appeared as the basis for a Roman mosaic and also for a Byzantine mosaic. The tiling of Figure 50(c) is found in a Roman pavement. The modern tiling of Figure 50(d) is of Arabic origin, as is also the tiling of Figure 50(e). All the tiles except the one of Figure 50(f) are constructed from semicircles and quadrants.

70° *A lamp base.* Just as there are curves filling the interior of a square (see Item 332° of *Mathematical Circles Revisited*), there are curves filling the interior of a cube. As in the planar cases, the spatial curves are defined as limits of sequences of polygonal (space) curves. Figure 51 shows the first two stages leading to a curve that fills a cubical box. The second one of these, fashioned from iron rods, would make an unusual base for a modern table lamp.

71° *A quilt.* At first glance, the quilt illustrated in Figure 52 seems to be quite random. Nevertheless, there is a certain symmetry in it and a certain reason behind its construction.

If we should think of the quilt as a chessboard, and each cell with

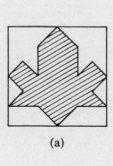

(a)

(b)

FIGURE 49

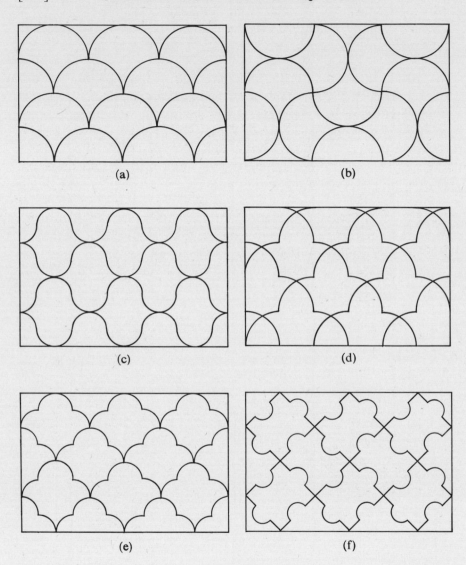

FIGURE 50

an R in it as containing a queen, we would have a solution of the eight queens problem (that is, we would have eight queens on the board so located that no queen can take any other queen). The same

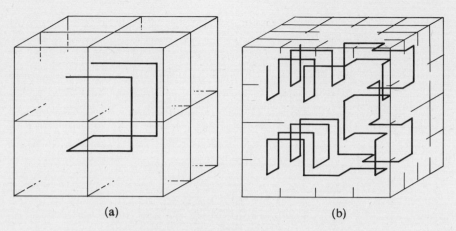

(a)   (b)

FIGURE 51

	P	R	O	B	G	Y	
M	B	Y	G	R	P	O	N
P	R	M	B	O	N	G	Y
G	Y	O	N	M	B	P	R
O	N	G	P	Y	R	M	B
Y	M	B	R	G	O	N	P
B	G	P	M	N	Y	R	O
R	O	N	Y	P	M	B	G

FIGURE 52

is true of each of the letters *B, Y, G, O,* and *P*—each leads to a solution of the eight queens problem. The letters *M* and *N* each lead to seven queens on the chessboard such that no queen can take any other. We thus have six superposed sets of eight queens along with two further

51

superposed sets of seven queens, filling sixty-two of the sixty-four squares of the chessboard. Thorold Gosset has shown that it is impossible to have eight superposed sets of eight queens.

It is to be noticed that the eight sets are, in pairs, reflections of one another in the vertical bisecting line of the board. Thus the R's and the G's, the B's and the O's, the Y's and the P's, and the M's and the N's are symmetrical arrangements. If the cells of each pair of symmetrical arrangements were to be colored alike, we would obtain the attractively symmetrical quilt of Figure 53.

Figure 53

72° *Some more quilts.* A *pentomino* is a planar arrangement of five unit squares joined along their edges. There are exactly twelve types of pentominoes; they are pictured in Figure 54.

An interesting puzzle is to put the twelve pentominoes together to form an 8 × 8 square with a square 2 × 2 hole in the middle. In 1958, Dana S. Scott instructed the MANIAC digital computer to search out all solutions to this puzzle. After operating for about three and a half hours, the machine produced a complete list of sixty-five distinct solutions, wherein no solution can be obtained from another by rotations and reflections. In every solution, the straight pentomino appears along an edge of the 8 × 8 square. There are exactly seven

Figure 54

solutions with no "crossroad," that is, with no point where four corners meet; Figure 55 shows one of these. There are a great many ways

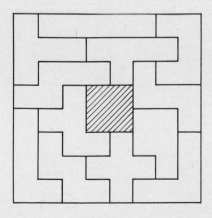

Figure 55

53

of forming the twelve pentominoes into an 8 × 8 square in which the four extra unit squares are separated but symmetrically arranged; Figures 56, 57, and 58 show three of these arrangements. Any one of the four designs illustrated in Figures 55, 56, 57, 58 could admirably serve as a quilt for a double bed. Figure 59 shows the twelve pentominoes formed into two 5 × 6 rectangles; these might serve for quilts on a pair of twin beds.

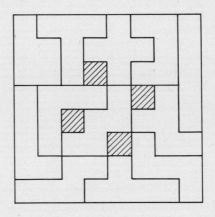

FIGURE 56

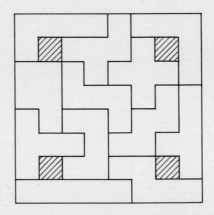

FIGURE 57

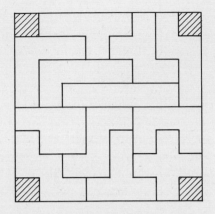

FIGURE 58

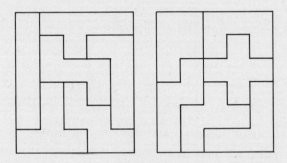

FIGURE 59

All twelve pentominoes can be assembled into a 6 × 10, a 5 × 12, a 4 × 15, or a 3 × 20 rectangle, but we leave these assemblages as puzzles for the reader.

GEOMETRY

HAVING just finished a discussion involving a number of attractive designs from geometry, it is perhaps fitting to complete the present quadrant with a few stories about the venerable science of geometry itself.

73° *An analogy.* The Euclidean foundation of geometry is to the Gaussian foundation of geometry as the Newton particle concept of physics is to the Faraday-Maxwell vector concept of physics.

74° *Another analogy.* The decomposition of a tensor into components appears to be an artificial act, for the components of the tensor do not belong to the essential substance of the tensor. Setting up a reference system to analyze a tensor by its components is like erecting a scaffold to study a building by its parts. The scaffold does not belong to the building, but it certainly fulfills a useful purpose.

75° *A science fiction possibility.* Some geometric concepts have been utilized for science fiction purposes, for example the Möbius strip, the tesseract, and higher-dimensional spaces have each been frequently used. An apparently overlooked possibility is the Riemannian sphere, a spherical surface in which antipodal points are identified. A peculiar feature of the Riemannian sphere is that upon circumnavigation and return to one's point of beginning, the directions left and right become interchanged. The navigator, on arriving at his home port, finds his directional sense confounded, for what was originally on his right now appears on his left, and the whole city has turned into its mirror image. He cannot even read the newspapers, for now the printing goes backward from right to left. But after one more circumnavigation, everything returns to familiar order, and the strange reversal of directions becomes rectified.*

76° *The truth of a theory.* In Book VII of *The Republic*, Plato discourses on the finding of the laws that control the movements of the

---

* For a fictional application of the somewhat similar idea of making coincide two symmetrical figures (like a right-hand and a left-hand glove) of a three-dimensional point space immersed in a four-dimensional point space, see *The Plattner Story* by H. G. Wells.

heavenly bodies, and he asks if one should look up to the sky to find these laws. He compares such observations to the visual scrutiny and measurement of a geometrical diagram, and he points out that no amount of such scrutiny and measurement, even of the most perfectly constructed diagram, will truly find the geometrical properties of the figure—these must be found by rigorous deduction from given basic principles. Astronomy, Plato concludes, should be pursued in a similar way, and therefore the astronomer who looks *down* has a better chance of finding the laws of the heavens than the astronomer who looks *up*. Newton and Einstein each became a Plato's astronomer and, by gazing down rather than up, found inner secrets of the heavens. "The truth of a theory," said Einstein, "is in your mind, not in your eyes."

77° *Zeuxis and Apollodorus.* The development of projective geometry in the seventeenth century and later owed much of its inspiration to art; earlier Renaissance artists and architects, seeking a way to produce truer representations, sought the laws of perspective. Thus artists like Leonardo da Vinci and Albrecht Dürer were fore-runners of the European development of projective geometry.

Now projective geometry did not originate with the Europeans, for we find certain aspects of the subject already considerably developed by the ancient Greeks. But here, too, the inspirational source in all likelihood lay in the art of the times. The Greeks excelled not only in philosophy and science, but equally well in the fine arts, as the sculpture and architecture of their day attest. In the fine arts, the Greeks were hardly less accomplished in the field of painting and drawing, but, because this medium is much less durable, time has erased almost all of their efforts, and to form some judgment of Greek skill in this area we must resort to literary allusions.

It seems that the Greek playwrights, vying with one another to produce superior plays, demanded props and painted backdrops. These were supplied by the artists, who, in turn, in order to produce naturalistic effects, required the basic laws of perspective. These laws, accompanied by a later technique of shading, led to some highly realistic backdrop paintings.

There is a legend claiming that the two greatest exponents of three-dimensional illusion through perspective and shading were the painters Zeuxis and Apollodorus (ca. 400 B.C.). These painters were in friendly rivalry with one another, and it was difficult to tell which was the greater master. Zeuxis once painted a bunch of grapes that appeared so realistic that birds came and picked at the canvas. A little later, Apollodorus invited Zeuxis to inspect his latest painting of a landscape. Zeuxis, on viewing the painting, asked his friend to pull aside the curtains at the two sides of the painting, so that he might view the whole canvas—but the curtains were a part of the painting. Zeuxis thereupon conceded superiority to his friend Apollodorus. He said, "Zeuxis deceived the birds, but Apollodorus deceived Zeuxis."

78° *Maps.* In bygone days a sailor found it convenient to set the course of his vessel at a fixed reading of his compass—constantly heading the prow of his vessel, for example, always in the direction NNW by his compass. Such a course intersects the earth's meridians at a constant angle and is called a *loxodrome*. To serve this method of sailing, the Renaissance geographer Gerhard Mercator (1512–1594) devised a method of mapping a sphere onto a plane in which loxodromes on the sphere map into straight lines on the plane. Such a mapping has become known as a *Mercator projection*, and a sailor using a Mercator projection could easily plot a course on his map simply by drawing the appropriate straight line path. He would then steer this course by holding the vessel at the corresponding fixed reading of the compass. Today the communication satellites make it very easy to pinpoint the position of a vessel at any time of day or night, and the once indispensable compass is no longer a vital instrument in ship navigation.

In air travel, a pilot generally wants to choose his course so as to arrive at his destination in the shortest time and with the least consumption of fuel. This means that he wishes to fly over great circle paths on the earth. An ideal map for the pilot would then be one which maps, not loxodromes of the sphere, but great circles of the sphere, into straight lines on the map, for on such a map the pilot could plot his desired route simply by drawing the straight line from his point of departure to his point of arrival. Maps of this sort, called *geodetic* *mappings*, are easily made; one simply projects the sphere from its center onto any appropriate tangent plane of the sphere.*

79° *An anthology of great poetry.* Apologizing for the many omissions in his elegant book, *Space through the Ages*, Cornelius Lanczos says, "Completeness could not be aimed at in a book of this nature; nor was it in any way contemplated. An anthology of great poetry must always compromise and be guilty of omissions. And so it is with this anthology of great geometrical ideas—so akin to an anthology of great poetry." †

80° *The method of dual languages.* Cornelius Lanczos, writing of the principle of duality of projective geometry in his book *Space through the Ages*, says, "One of the most surprising phenomena of projective geometry is the point-line dualism. Let us imagine that somebody studies projective geometry from a textbook written in a foreign tongue. Owing to an oversight he translates two words wrongly. What is meant by the word 'point' is translated as 'line,' and *vice versa*. Then the peculiar fact holds that he is not able to discover his mistake because all statements remain correct, in spite of the different interpretation given to them." †

81° *Blind geometers.* The invention of analytic geometry was a development that changed the complexion of geometrical research. Unforeseen advances became possible because of the abstract nature of

---

* The Mercator projection maps the point with latitude θ and longitude ϕ of the sphere into the point on the plane having cartesian coordinates (x,y) given by

$$x = \phi, \quad y = - \log \tan \theta/2.$$

The geodetic mapping of the northern hemisphere of the earth onto the tangent plane at the earth's north pole maps the point with latitude θ and longitude ϕ into the point of the plane having polar coordinates $(r, \alpha^2$ given by

$$\alpha = \phi, \quad r = \tan \theta.$$

† Quoted by permission of Academic Press, New York.

algebraic thinking, but also, in a sense, the esthetic satisfaction derived from the contemplation of geometric figures was taken away. One no longer had to "see" what happens in space. Even a blind person, who has never seen a straight line, or a circle, or an ellipse, or a sphere, can investigate the properties of these figures. Geometrical invention became a manipulative and routine procedure, and geometry became open to anyone—even to one possessing no trace of a space perception. In a way we are all like this blind geometer when we study geometry in higher dimensions than three.

82° *The true meaning of the Delphian oracle.*   Plato was born in or near Athens in 427 B.C., probably the year in which a severe plague in Greece killed off a large portion of the population of Athens. A story is told that a delegation of citizens of Delos was sent to the oracle of Apollo at Delphi to find out how the gods might be placated and the plague ended. The oracle explained that the gods were dissatisfied with the size of the cubical altar to Apollo and that the altar must be doubled in size. This led to the problem of constructing the edge of a cube having twice the volume of a given cube. In spite of assiduous effort, the Greeks were unable to obtain a theoretically exact solution to this problem using compasses and straightedge.

Now it is difficult to believe that the gods would not be satisfied by a reasonably close approximation to the doubled altar, but to the Greek mind the theoretically exact solution of the problem became one of eminent scientific interest. When Plato was asked as to the true meaning of the reply made by the oracle, he gave a characteristic answer. He said he believed the reply was designed to point out the shameful neglect on the part of the Greeks of the scientific pursuit of geometry and to restimulate interest in pure geometrical research.

83° *The origin of geometry.*   Many historians have advanced the thesis that the beginnings of geometry are to be found in the early surveying practices of the ancient Egyptian rope-stretchers; indeed, the word "geometry" means "measurement of the earth." Another thesis is that geometry had its origins in religious ritual and the construction of holy altars. A third thesis can be defended—namely that the beginnings of geometry are to be found in the stars.

Mesopotamian shepherds in their lonely night vigils could not but help observe, in the clear Babylonian skies, the courses of the moon and the stars, and to notice the ever-returning constellations. The regularity of the motions of the sun, moon, and planets evoked in early man a desire to understand something of the mystery of these astronomical events, and to predict their recurrences. Thus primitive astronomy cradled the natural sciences and furnished the starting point of geometrical theories. The concept of a "point" is suggested by the stars themselves. Triangles, quadrilaterals, and other rectilinear figures appear in the constellations. The circle is seen in the periphery of the sun and the moon. The misty beginnings of geometrical thinking may have arisen in this purely subconscious fashion through simple star-gazing. This thesis is advanced by Cornelius Lanczos in his beautiful book, *Space through the Ages*.

84° *Geometers and analysts*. The story on p. 85, Vol. 2, of *In Mathematical Circles*, about botanists being nice and mathematicians being nasty reminds me of Jesse Douglas who used to say that geometers were nice and analysts nasty.

And in Moscow, in the thirties, I was told that early in the century the liberal students at the University went to hear Kagan, the geometer (who was a Jew), and the reactionary ones went to Egorov, the analyst (who was Orthodox and an anti-Semite).—DIRK J. STRUIK

85° *DIN*. The Deutsche Industrie Normen (DIN), a German committee on standardization, recommended that the shapes and sizes of the sheets of paper used for printing, typewriting, letter writing, note pads, and so on, be chosen so as to minimize waste in cutting the larger sheets to the smaller sizes. This economy of paper is achieved by choosing the proportions of a sheet so that when the sheet is halved the two smaller sheets will have the same shape as the original sheet. Thus a sheet of typing paper, when halved, could produce two smaller, similar sheets suitable for personal letter writing, and each of these, when halved, could produce two still smaller, similar sheets for pages of a note pad. If the length and width of the standard sheet are x and y units, then, after one halving,

$$x:y = y:x/2,$$

whence $x = y\sqrt{2}$.

Since $\sqrt{2}$ fairly closely approximates the golden ratio $(1 + \sqrt{5})/2$, it follows that the rectangle which is most economical is at the same time the most pleasing, for it does not differ seriously from the golden rectangle which artists find so congenial.

86° *An early love affair.* Bertrand Russell (1872–1970) began the study of Euclid's *Elements* when he was eleven, with his eighteen-year-old brother serving as tutor. In his *Autobiography*, Russell says, "This was one of the great events of my life, as dazzling as first love. I had not imagined there was anything so delicious in the world. . . . From that moment until . . . I was thirty-eight, mathematics was my chief interest and my chief source of happiness."

87° *Albrecht Dürer's approximate constructions.* Although an arbitrary angle cannot be trisected exactly with compass and straightedge, there are constructions with these tools that give remarkably good approximate trisections. An excellent example is the construction given in 1525 by the famous engraver and painter Albrecht Dürer. Take the given angle *AOB* as a central angle of a circle (see Figure 60). Let *C* be that trisection point of the chord *AB* which is nearer to *B* (this can be found exactly with compass and straightedge). At *C* erect the perpendicular to *AB* to cut the circle in *D*. With *B* as center and *BD* as radius draw an arc to cut *AB* in *E*. Let *F* be the trisection point

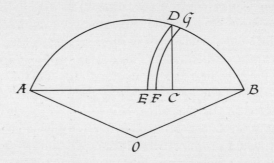

FIGURE 60

of *EC* which is nearer to *E*. Again, with *B* as center, and *BF* as radius, draw an arc to cut the circle in *G*. Then *OG* is an approximate trisecting line of angle *AOB*. It can be shown that the error in trisection increases with the size of the angle *AOB*, but is only about 1″ for angle *AOB* = 60° and about 18″ for angle *AOB* = 90°.

An approximate construction, given by Albrecht Dürer, of a regular nonagon inscribed in a given circle of radius *r* is as follows (see Figure 61). Draw a circumference concentric with the given circle and having its radius equal to 3*r*. Divide this circumference into six equal parts by the points *A, B, C, D, E, F*. With *F* and *B* as centers, draw arcs through *A* and the common center *O* of the two circles, these arcs cutting the given circle in *M* and *N*, respectively. Then

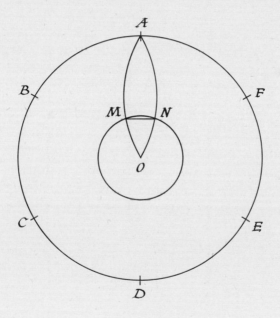

FIGURE 61

MN is approximately a side of the sought nonagon. The approximation, this time, is not very good. A considerably better approximation is to take half of angle *NOB* as the central angle of the required nonagon.

63

88° *Two daffynitions.* *Topology* is the study of badly drawn figures (illustrated by the topological circle shown in Figure 62). A *topologist* is a person who doesn't know the difference between a doughnut and a coffee cup (illustrated in Figure 63).

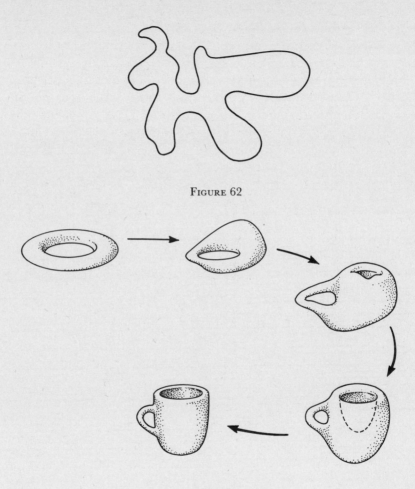

FIGURE 62

FIGURE 63

89° *The birthplace of topology.* Sometime in the first quarter of the eighteenth century a gala affair was planned in old Königsberg of Prussia. Part of the festivities was to be a parade that would march in a circuitous course about the town. Now Königsberg was on the Pregel River, about four and a half miles from where the river empties into the Baltic Sea and where there is a small island called the Kneiphof (beer garden), inasmuch as it lacked a water supply of its own. There were seven distinctive bridges, each with its own cherished design, connecting various banks of the Pregel River (see Figure 64). The cathedral of Königsberg was on the island, and adjacent to it the university and the grave of Königsberg's best known son, the famous philosopher Immanuel Kant. It was hoped to start the parade at the

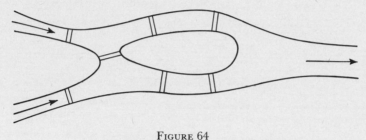

FIGURE 64

cathedral on the island, cross each of the bridges once and only once, and return to the cathedral on the island. But try as they would, no one was able to plot a course of the desired sort.

In time the problem reached the great Swiss mathematician Leonhard Euler, who, in 1736, submitted to the St. Petersburg Academy a paper on linear graphs in which the general problem of any number of islands and bridges was considered. In the paper, the original Königsberg bridge problem was shown to be impossible, and the subject of topology received its start.

Today Königsberg is known as Kaliningrad and the original arrangement of bridges has been augmented. The old academic atmosphere is gone and a military one has replaced it, for Kaliningrad is the largest naval base of the Soviet Union. The author of the present book once possessed, in MMM*, a beautiful aerial view of Königsberg

---

* MMM (My Mathematical Museum). See Items 246°, 247°, 248°, and 249° of *Mathematical Circles Revisited*.

and its seven famous bridges taken from the Graf Zeppelin. One wonders where the picture is now.

90° *Q. E. D.* The custom arose in olden days of appending the letters Q. E. D. (standing for "quod erat demonstrandum," meaning "that which was to be proved") at the end of the proof of a geometric theorem. Today, in the geometry classroom, one might put Q. E. D. at the end of a particularly simple and straightforward proof, and tell the class that it there stands for "quite easily done." Again, the same letters might be used in the geometry class at the end of a particularly beautiful and clever proof, and the class told that the letters now stand for "quite elegantly done."

QUADRANT TWO

From super-respectability

to a mathematical quiddity

FROM THE AMERICAN SCENE

In this section we give a few stories about some American mathematicians: William Fogg Osgood, the great analyst at Harvard University during the late eighteen and the early nineteen hundreds; Edward Kasner, the brilliant geometer of Columbia University in the first half of this century; Albert Einstein, world-famous mathematical physicist who, though born in Germany, is quite properly claimed by America; Robert Lee Moore, America's first topologist; Norbert Wiener, the M.I.T. genius and eccentric; Spofford Kimball, who was for over a quarter of a century the beloved Chairman of the Mathematics Department of the University of Maine. The section is concluded with five stories connected with the Problem Department of the *American Mathematical Monthly*, with which the present writer was editorially associated for over twenty-five years. It would be easy to devote a whole volume just to stories and anecdotes about American mathematics and mathematicians.

91° *Super-respectability.* It was said at Harvard that Professor William Fogg Osgood (1864–1943) was so respectable and took himself so seriously that every morning, when he got up, he looked at himself in the mirror and said, "Good morning, Professor Osgood."

Dirk J. Struik

92° *Practical rambler.* Clifton Fadiman has told some amusing anecdotes about Edward Kasner (1878–1955),* the charmingly eccentric one-time Adrian Professor of Mathematics at Columbia University. This, and the next three items, are examples.

Kasner was fond of solitary rambling, particularly along the Palisades in New Jersey and New York. To save himself the discomfort of toting too much eating paraphernalia, he carried only his tea and sugar, and had the other items, such as teapots and essentially im-

---

* Clifton Fadiman dedicated his delightful anthology, *The Mathematical Magpie* (Simon and Schuster, 1962), to the memory of Edward Kasner.

perishable foods, cached in a series of lunch boxes hidden in secret places along his route.

93° *Mountain climber.*   Kasner spent his summer vacations in Brussels, and each summer he organized a mountain-climbing expedition to the highest point in Belgium. He claimed he experienced great satisfaction in yearly attaining this peak. When asked the height of the eminence, he would reply, "Twelve feet above sea level."

94° *Creator of suspense.*   Kasner appeared never to wear any but a loose pepper-and-salt suit. Since he long ago abandoned the belt as an extraneous modern invention, he spent a lot of time holding up his trousers, and during his lectures there would be much speculation as whether he would or would not make it to the end before his pants would fall down.

95° *Epicurean.*   Kasner maintained that the best food in New York City was to be found in one of the city's Automats. He would occasionally visit a first-class restaurant solely to reassure himself of his opinion.

96° *Einstein and religion.*   Albert Einstein (1879–1955) was deeply impressed by the manner in which cosmic happenings obey rational and universal laws. The operation of these laws, with their rational motivation, appeared to him as proof of a cosmic wisdom that he called God. On the other hand, he could not accept the humanized qualities of a personal God, such as a God of retribution and punishment, or a God of love and a listener to the supplicant, and so on, which seems to be demanded by the codified religions. Such attributes, he felt, were perhaps introduced by a misunderstanding or misinterpretation of the prophetic teachings. He accordingly called himself a "deeply religious unbeliever."

97° *A sufficient reward.*   It seems that a little girl in Princeton had trouble with her homework in arithmetic. Suddenly, her homework improved immensely. The teacher, amazed, asked how she had made such progress. "Oh," said the girl, "there is a nice old gentleman on

Mercer Street who is helping me." The teacher thought that it was best to tell the mother what was happening. This lady, embarrassed no end, went to see the nice old gentleman. "But Professor Einstein, why do you spend your time helping my little girl? What do you get out of it?" Said Einstein, "But I do get something out of it. Every time I help her she gives me a lollipop."—DIRK J. STRUIK

98° *Einstein and his check.*   It has been told that when Dr. Einstein received his first check as a member of the Institute of Advanced Study at Princeton, New Jersey, he used the back of the check for some mathematical figuring and then in a short time lost the scrap of paper. After that, his stipend checks were turned over directly to Mrs. Einstein for safer keeping.

99° *The prankster.* Robert Lee Moore indulged in innocent pranks. For example, it is said that when inviting some of the bright young graduate students of mathematics to his home for dinner, he would afterwards measure their heights with a yardstick that was really only thirty-five inches long.

100° *Identification.*   Norbert Wiener (1894–1964) tells, in his book *I Am a Mathematician,* how the mathematicians George Bouligand and Leon Lichtenstein identified themselves when he met each of them for the first time. Bouligand was awaiting Wiener's arrival at the Portiers railway station, prominently holding up a copy of one of Wiener's articles. Later, in Germany, Lichtenstein similarly identified himself at the railway station by holding up a sheet of paper on which he had written, in Wiener's honor, the principal formula of potential theory. Leon Lichtenstein and Norbert Wiener were cousins.

101° *How Norbert Wiener got his beard.*   It was in the summer of about 1927 that Norbert Wiener and Dirk Struik undertook an extended hiking expedition into the Presidential Range of New Hampshire, from which they returned with luxurious growths of beard. Struik, who Norbert claimed now resembled a Rembrandt painting, soon shaved off his beard, but Mrs. Wiener trimmed Norbert's down to the more modest proportions that he continued to wear the rest of his life.

102° *Wienerian wit.* A large group of Americans once attended a mathematical gathering in Guadalajara, Mexico. Among the Americans was Professor Francis Murnaghan of Johns Hopkins University, who experienced considerable difficulty adjusting to the Mexican diet. One morning Arturo Rosenblueth said, in English and to no one in particular, "I feel fine, period!" Norbert Wiener, who was standing nearby, replied, "Murnaghan feels rotten, colon."

103° *The modern physicist.* Describing the state of present-day physics, Norbert Wiener said, "The modern physicist is a quantum theorist on Monday, Wednesday, and Friday and a student of gravitational relativity theory on Tuesday, Thursday, and Saturday. On Sunday he is neither, but is praying to his God that someone, preferably himself, will find the reconciliation between the two views."

104° *Shop-talker.* When Heisenberg was at M.I.T., in the thirties, Wiener was constantly visiting him, talking mathematics. Said Heisenberg to some of us, "Don't you think that Professor Wiener takes mathematics a little too seriously?"—DIRK J. STRUIK

105° *Wienerian disease.* After the publication of his *Cybernetics*, Wiener was often embarrassed by the enthusiasm of some of his followers, who saw in it a kind of universal panacea. Remembering that Marx had exclaimed that he was not a Marxian, Wiener said to one of his friends, "I am not a Wienerian." Answered his friend, "Yes, but there are plenty with Wienerian disease."—DIRK J. STRUIK

106° *An incomplete story.* The story of Item 358° of *In Mathematical Circles* is incomplete. Wiener came from the cafeteria and met and talked with the student. After the talk was over, Wiener asked, "Do you recall, when we met, in which direction I was headed?" The student told him. "Ah," said Wiener, "then I have had my luncheon."
DIRK J. STRUIK

107° *How a young mathematician saved himself by ingenuity.* Norbert Wiener has told an amusing anecdote about one of the instructors in the Mathematics Department of M.I.T. This young man later

went to Munich to study and there got himself entangled in a challenge
to duel an army officer. To escape the duel, he used his right to name the
weapons. He chose bows and arrows and then had some of his friends
adroitly noise it around that he was a champion archer. When some
of his colleagues at M.I.T. read the story in the Paris edition of *The
New York Times*, they sent the young man a letter purporting to come
from an archery society and offering him the highest honor the
society could bestow.

108° *Crackpots.*   Professor Spofford Kimball, for many years the
Chairman of the Mathematics Department of the University of Maine
at Orono, had a way of discouraging angle trisecters, cube duplicaters,
and circle squarers. When he received a supposed construction from
some crackpot, he politely replied that his professional fee for examining
the construction would be $100. This usually terminated the bother-
some correspondence.

109° *Benjamin Franklin Finkel.*   Since its inception in 1894, *The
American Mathematical Monthly* has been, among other things, a problems
journal. Benjamin Franklin Finkel, the founder, was himself an en-
thusiastic problemist and was the author of the once popular *Finkel's
Mathematical Solution Book*.

Professor Finkel liked to tell the story of how, at the age of fifteen,
while attending a small country school in Ohio, his interest in mathematics was aroused by a senseless problem that was going the rounds of
the country. An older half brother of his, hearing the problem discussed at a village grocery store gathering, brought it home for him
to think on. He took it to his schoolteacher who pointed out how it could
possibly be solved by geometry. Having never seen a geometry book,
this gave Benjamin little satisfaction, and he tried to apply the rules of
mensuration found in his copy of Ray's *Third Part Arithmetic*. His
efforts were crowned with success several years later. The problem
concerned was: "There is a ball 12 feet in diameter on top of a pole 60
feet high. On the ball stands a man whose eye is 6 feet above the ball.
How much ground beneath the ball is invisible to him?"

This problem later found its way into *Finkel's Mathematical Solution
Book,* and it so impressed Finkel in connection with his own introduction
to mathematics that throughout his life he advocated the educational

value of a good senseless problem as against that of the so-called practical problem which is so insisted upon by some modern educators.

When Finkel's plan to start the publication of *The American Mathematical Monthly* was about to crystalize, he looked about for an associate, and finally invited Professor John M. Colaw of Monterey, Virginia, to join him. He had never personally met Professor Colaw, but had been impressed by the man's contributions, mostly problems, to *The School Visitor*, to which Finkel was an avid subscriber.

With the personal interest of the two associates in problems, and the difficulty of a new journal at that time to obtain high grade articles by top men, it is little wonder that for the first few years *The American Mathematical Monthly* was predominantly a problems journal, and bore on its title page the following description of purpose: "Devoted to the solution of problems in pure and applied mathematics, papers on mathematical subjects, biographies of noted mathematicians, etc."

In the introduction to the first issue of *The American Mathematical Monthly* Finkel said, "The solution of problems is one of the lowest forms of mathematical research, . . . yet its educational value cannot be overestimated. It is the ladder by which the mind ascends into higher fields of original research and investigation. Many dormant minds have been aroused into activity through the mastery of a single problem." One can take a stronger stand: It is by means of problems that mathematics develops and actually lifts itself by its own bootstraps. Every research article, every doctoral thesis, every new discovery in mathematics, results from an attempt to solve some problem. Not only is the posing of appropriate problems a very suitable way to introduce the promising student to mathematical research, but the ability to propose significant problems is one requirement for being a creative mathematician.

110° *Problemist extraordinaire.* When *The American Mathematical Monthly* started in 1894, the problem section of the journal immediately attracted G. M. B. Zerr, a professor and school administrator, chemical engineer, and highly versatile mathematician. In the first seventeen years of the journal's existence, G. M. B. Zerr, right up to his untimely death in 1910, far surpassed all other problemists in both the quantity and the quality of his contributions. In those seventeen years he amassed the incredible total of 1697 contributions, and during this same period he contributed with equal fervor to the problem sections of a number of other contemporary magazines.

111° *Two prizes.* Two of the problems in *The American Mathematical Monthly* had prizes attached to them. The earlier of these appeared in the very first volume of the journal, as Algebra Problem 43, p. 433 (1894). F. M. Shields of Cooperwood, Mississippi, proposed a complicated time-rate-distance problem and offered "a city lot at St. Andrews, Florida, to the party sending the EDITOR the first correct answer." The prize was collected by G. B. M. Zerr (see Item 110°).

The second problem appeared in the December 1929 issue of the journal. S. A. Corey of Des Moines, Iowa, in proposing Problem 3402, offered a prize of $10 "to the person sending in a solution or comment most enlightening, in the judgment of the editors." This problem is still in the list of unsolved problems of the journal, and so the prize has remained uncollected.

112° *A student from Missouri.* Professor L. M. Kelly, when teaching a freshman course in analytic geometry at the University of Missouri in about 1946, once casually remarked to his class that if the inside of a race track is elliptical, and the track is of constant width, then the outside of the track is not necessarily an ellipse. Among the students in the class was one truly "from Missouri," who asked to be shown that the outside need not be an ellipse. Now this was not the easiest thing to do at the student's level of mathematical understanding. As a result, Professor Kelly proposed, as Problem E 753 in the January 1947 issue of *The American Mathematical Monthly*, the question, "How can one convince a class in elementary analytics that if the inside of a race track is a noncircular ellipse, and the track is of constant width, then the outside is not an ellipse?" Solutions, which may have been beyond the student, appeared in the September 1947 issue of the journal.

113° *Cupid's problem.* In the October 1946 issue of *The American Mathematical Monthly* appeared the attractive little problem E 740: "Let there be given five points in the plane. Prove that we can select four of them which determine a convex quadrilateral." This problem was published as proposed by Esther Szekeres, then of Shanghai, China, and when solutions to the problem were published in the May 1947 issue of the journal, the problem was titled "Cupid's problem." There is no doubt that very few readers of the *Monthly* understood why the problem received this particular title. Actually, the problem had been submitted, not by Esther Szekeres, but by a close friend of both Esther and her husband George Szekeres. The friend, who wished to remain anonymous, stated that it was this problem and some of its generalizations* that first brought Esther and George together, and he asked that the problem be published as proposed by Esther. Cognizant of the above romantic facts, the editor of the Problem Department, when solutions were published, titled the problem "Cupid's problem."

AMONG THE ENGLISH

HERE we offer a score of stories about an assortment of English mathematicians from John Blissard, born in 1803, to J. E. Littlewood, born in 1885. The others represented are Augustus De Morgan, J. J. Sylvester, Isaac Todhunter, William Kingdon Clifford, Sir Arthur Eddington, and William Sealy Gosset. Mathematicians William Frend, A. H. Frost, Roger Cotes, Isaac Newton, Robert Tucker, G. H. Hardy, Alexander Macfarlane, Paul Halmos, Oswald Veblen, and Norbert Wiener come in for incidental mention—all English except the last four. Tales have already been told about a number of the above men in our two earlier collections of mathematical stories and anecdotes.

114° *Blissard's sermon.*   John Blissard (the last name is pronounced with the accent on the second syllable) is noted in the history

---

* The general case of this problem is discussed by Paul Erdös and George Szekeres in an article in *Compositio Mathematica*, vol. 2 (1935), pp. 463–470. From nine points in the plane we can select five which determine a convex pentagon, but it was not known if from $2^{n-2} + 1$ points in the plane we can select n which determine a convex n-gon. Given $n + 3$ points in an n-flat, we can select $n + 2$ which determine a convex polytope.

of mathematics almost solely for his "symbolic method" or "umbral calculus." He was born in 1803 in Northampton, England, where his father was a physician, and he died in 1875 in the inconspicuous village of Hampstead Norreys, where he conscientiously served as a clergyman for forty-six years. Shelved at Hampstead Norreys, where he and his wife reared twelve children, his mathematical talents were all but buried. Nevertheless, he was never heard to express any discontent with his lot, and he managed to keep his mathematical soul alive by coaching a few private pupils for the Cambridge mathematics examinations, and by ruminating on mathematical problems.

It is said that the schism in Blissard's loyalties to his parish duties and to his mathematical proclivities led him to become absentminded as he aged. The following anecdote is told about this amiable clergyman. In those days, churchgoers felt shortchanged if the sermon was not of epic proportions, and one Sunday Blissard unwittingly satisfied his parishioners in this respect. He chose as his text the twenty-third psalm, each sentence of which was to serve as a subtext for a subsermon within the overall delivery. When he reached "He restoreth my soul," the subsermon attained such an extraordinary length that he forget where he had left off, and gave forth once more with "He restoreth my soul," and delivered another lengthy sermon on this same subtext completely different from his first one. At the conclusion he again lost his cue and for a third time came out with "He restoresth my soul," and delivered a third and yet different subsermon on these words. "All three addresses," said his wife later, "were quite different and all quite perfect each in its own way. And the people didn't mind a bit."

115° *From Blissard's obituary.*   When John Blissard died in 1875, the *Reading Mercury, Berks County Paper*, recorded, "Throughout his life he [Blissard] scarcely made an enemy, and although his house was once set on fire by an evil servant he forgave her, when she admitted her guilt."

116° *Unattached.*   Augustus De Morgan (1806–1871) was born in India, and his father and grandfather had also been born there. Because of this, De Morgan used to say he was neither English, Scottish, nor Irish, but was a Briton "unattached," using the term technically

77

as applied to an undergraduate of Oxford or Cambridge who is not a member of one of the colleges of his university.

117° *A man of peculiarities.*   When Augustus De Morgan's friend, Lord Brougham, was installed as Rector of the University of Edinburgh, the University Senate offered to confer the honorary degree LL.D. on De Morgan. De Morgan declined the honor, claiming that as applied to him the degree was a misnomer. He once signed his name as:

Augustus De Morgan

H·O·M·O · P·A·U·C·A·R·U·M · L·I·T·E·R·A·R·U·M

De Morgan did not enjoy the country, and when his family was vacationing at the seaside, or the British Association was meeting in the countryside, he stayed behind in the heat and dust of the city. He said he felt like Socrates, who once declared that the farther he got from Athens the farther he was from happiness.

De Morgan never attended a meeting of the Royal Society and never sought membership in the Society. He never voted at an election, and he took pride in the fact that he never visited the House of Commons, the London Tower, or Westminster Abbey.

118° *Mrs. De Morgan.*   When Augustus De Morgan took up residence in London he found a congenial friend in William Frend, a one-time second wrangler and a fellow arithmetician and actuary sharing similar religious views. In 1796 Frend had published a book, *Principles of Algebra,* protesting against the use of negative numbers. Frend lived in the outskirts of London in a country house that had earlier been occupied by the famous writer Daniel Defoe and the great preacher and hymnologist Isaac Watts.

De Morgan, who performed exquisitely on the flute and was always merry company, was a frequent and very welcome visitor at the Frends' home. The outcome was that in 1837, when he was thirty-one years old, De Morgan married Sophia Elizabeth, one of Frend's daughters. Sophia was interested in spiritualism, and later, as Mrs. De Morgan, wrote a book describing such phenomena of spiritualism as table rapping, table turning, and voices from the grave. She also put her husband's essentially completed and highly entertaining work, *A Budget of Paradoxes*, through the press after he had died.

119° *Definite integrals.* At one period in the history of mathematics it was very fashionable to add to the list of interesting definite integrals, such as:

1. $$\int_0^\infty x^2 e^{-x^2}\, dx = \sqrt{\pi}/4,$$

2. $$\int_0^\infty dx/(1+x)\sqrt{x} = \pi,$$

3. $$\int_0^\infty (\cos x\, dx)/x = \infty,$$

4. $$\int_0^\infty (\tan x\, dx)/x = \pi/2,$$

5. $$\int_0^\infty (\sin^4 x\, dx)/x^4 = \pi/3,$$

6. $$\int_0^\infty \cos(x^2)\, dx = \sqrt{(\pi/2)}/2,$$

7. $$\int_0^1 x \log(1+x)\, dx = 1/4,$$

8. $$\int_0^1 (\log x\, dx)/(1+x) = -\pi^2/12,$$

9. $$\int_0^{\pi/2} (\log \tan x)\, dx = 0.$$

Collections of mathematical tables contain pages and pages of such integrals. Commenting on the fashion, J. J. Sylvester said, "It seems to be expected of every pilgrim up the slopes of the mathematical Parnasus; that he will at some point or other of his journey set down and invent a definite integral or two towards the increase of the common stock."

120° *No ear for music.*   Isaac Todhunter (1820–1884), the British mathematician who wrote so many fine textbooks and comprehensive histories of special areas of mathematics, had no ear whatever for music. He claimed he knew only two tunes—one was "God save the Queen," and the other wasn't. He said he recognized the former by the people standing up.

121° *A legend.*   Alexander Macfarlane has said that the following legend was applied to Isaac Todhunter, if not actually recorded of him. "Once on a time, a senior wrangler gave a wine party to celebrate his triumph. Six guests took their seats round the table. Turning the key in the door, he placed one bottle of wine on the table asseverating with unction, 'None of you will leave this room while a single drop remains.'"

122° *Mathematical talent foreshadowed.*   There is interest in knowing if a great mathematician, other than an obvious prodigy, showed any inclination for his subject at a young age. There are a couple of foreshadowing stories told of William Kingdon Clifford's young days. For example, when he was six years old his parents left him for a short time under the charge of his aunt Mrs. McLoed. Putting Willie to bed one night, the aunt found the boy looking very thoughtful, and she asked him what was on his mind. Smiling, the boy looked up at his aunt and said, "Aunt Annie, I don't think you would know." Upon pressing further, the little boy said he was calculating "the number of penknife edges it would take to go round the wheel of a coach." He gave his aunt the result of his calculation, along with the size of the wheel. The problem and answer were later submitted to an uncle of Willie's, Mr. Frank Kingdon, who was a fair mathematician. The calculation was found to be correct to within a few knife edges.

123° *Kite flying.*   William Kingdon Clifford remained much of a boy during all thirty-four years of his brief life, and he never cared to give up the games and play of his boyhood. Among the entertaining activities of his boyhood was that of flying kites. In 1863, when he was eighteen years old, Clifford wrote to a Mr. Miller "I have had in my mind, almost from the time I began to fly kites (I have not yet left off), the problem of finding the form of a kite string under the action of the wind. On a rough trial, the other day, the intrinsic equation seemed not very difficult to obtain; if I get any result, I will send it to you hereafter." The question of the kite string finally figured as Problem 6009 in the *Educational Times*, July 1879 and May 1880. Clifford died March 3, 1879.

124° *Unlocking a puzzle.* The Reverend Percival Frost once boasted to his brother A. H. Frost (who had done missionary work in India) of the remarkable space perception possessed by the young William Kingdon Clifford. Now the brother had brought back with him from India a complicated three-dimensional puzzle in the form of a sphere composed of a number of cleverly interlocked pieces, and the challenge of the puzzle was to take it apart. A. H. could scarcely believe the incredible things Percival said about the boy Clifford, and so he asked Percival to invite the youngster to tea and see if he could unlock the three-dimensional puzzle. Forthwith the boy was invited and shown the puzzle. Without touching the puzzle, Clifford carefully looked it over for a few minutes, and then sat with his head in his hands for a few more minutes. He then picked up the puzzle and, to A. H.'s astonishment, immediately took it apart.

125° *A repeated tribute.* The two very promising mathematicians, Roger Cotes (1682–1716) and William Kingdon Clifford (1845–1879), each died at the young age of thirty-four. Of Cotes, Isaac Newton once said, "If Cotes had lived, we had known something" (see Item 220° of *In Mathematical Circles*). When in 1882 Robert Tucker edited the mathematical papers of Clifford, he placed on the title page of the work the words, "If he had lived we might have known something."

126° *Eddington on the fourth dimension.* Sir Arthur Eddington, in his *Space, Time, and Gravitation*, comments on the concept of the fourth dimension as follows:

> However successful the theory of a four-dimensional world may be, it is difficult to ignore a voice inside us which whispers: "At the back of your mind, you know that a fourth dimension is all nonsense." I fancy that voice must often have had a busy time in the past history of physics. What nonsense to say that this solid table on which I am

writing is a collection of electrons moving with prodigious speed in empty spaces, which relatively to electronic dimensions are as wide as the spaces between the planets in the solar system! What nonsense to say that the thin air is trying to crush my body with a load of 14 lbs. to the square inch! What nonsense that the star cluster which I see through the telescope, obviously there *now*, is a glimpse into a past age 50,000 years ago! Let us not be beguiled by this voice. It is discredited. . . .

We have found a strange footprint on the shores of the unknown. We have devised profound theories, one after another, to account for its origins. At last, we have succeeded in reconstructing the creature that made the footprint. And lo! It is our own.

127° *Student.*   The English statistician William Sealy Gosset published his pioneer work on the theory of small samples under the name of "Student," perhaps, as Paul Halmos once remarked, to avoid embarrassing his employers, the well-known brewers of Guinness.

128° *Preparing for an examination.*   J. E. Littlewood, in writing of his mathematical education, tells of an experiment in studying that he once performed. He had an examination to prepare for over an Easter vacation, and he decided to bury himself at Hartland Quay, noted for its superb scenery and remoteness from all railway stations. He was to give up smoking and immerse himself in concentrated study of mathematics, relaxing—over poetry and philosophy—only in the evenings. The window of his room opened on the sea, and on arrival he ceremoniously tossed his pipes and tobacco into the water. The very next day he relapsed from his rigorous program, and thenceforth did only a negligible amount of studying. The experiment taught him that for serious work one seems to do best in a familiar setting with customary routine.

129° *Littlewood and the Dedekind cut.*   In a Dedekind cut, the set of all rational numbers falls into two classes, L and R, such that L has no greatest member and every number of L is less than every number of R. It was Littlewood who first introduced the logical letters L (for left) and R (for right) for the two classes, and he rightfully regarded this small contribution of notation as one for which students would be grateful.

In the first edition of Hardy's *Pure Mathematics*, the letters T and U

are used for the two classes; later editions use L and R. In the later editions Hardy gives a number of fine acknowledgments to Littlewood, but none in connection with the letters L and R. When Littlewood teasingly pointed out to Hardy that this contribution should also be acknowledged, Hardy refused on the ground that it would be insulting to mention anything so minor. Littlewood claimed this was merely the familiar response of the oppressor: "What the victim wants is not in his own best interests."

130° *Littlewood and Hardy.* Littlewood read the first proof-sheets of an article by Hardy on the remarkable Hindu mathematician Ramanujan. In the article Hardy had written, "As someone said, each of the positive integers was one of his personal friends." On reading this, Littlewood commented, "I wonder who said that; I wish I had." In the final proof-sheets Littlewood read (and this is how it came out in the printed article), "It was Littlewood who said, each of the positive integers was one of his personal friends."

131° *Littlewood and Veblen.* At one time Oswald Veblen gave a sequence of three lectures on *Geometry of paths*. At the conclusion of one of the lectures, the paths had worked themselves into the form

$$\frac{x-a}{l} = \frac{y-b}{m} = \frac{z-c}{n} = \frac{t-d}{p}.$$

Here Veblen broke off, briefly announced what was to come, and ended up with the words, "I am acting as my own John the Baptist." To this Littlewood (who was attending the lectures) commented, "Having made your own paths straight."

132° *A comment by J. E. Littlewood.* "A good mathematical joke is better, and better mathematics, than a dozen mediocre papers."

133° *Littlewood demonstrates mountain-climbing techniques.* A surprising number of American and British mathematicians are or have been amateur mountain climbers. J. E. Littlewood was one of these, and Norbert Wiener, in his book *I Am a Mathematician*, says that when he visited England in 1932, Littlewood was at the peak of his climbing career. Littlewood used to invite Wiener to his digs at Trinity and there demonstrate rock-climbing maneuvers by taking traverses about the pillars in Neville's Court.

TWO IRISHMEN

WILLIAM Rowan Hamilton (1805–1865) is without question the greatest mathematician Ireland has so far produced. Which Irishman should be accorded second place may not be so easily determined, but surely Henry John Stephen Smith (1826–1883) would be among the candidates. We have already (in *In Mathematical Circles* and in *Mathematical Circles Revisited*) told some seven stories about Hamilton, and Hamiltonian lore is so rich and extensive that we here tell some seven more tales about this eminent mathematician. While we are concerned with Ireland, we also tell three stories about Smith—a man not so rich in legend and stories as Hamilton, but whose neglect in our earlier collections should certainly now be rectified.

134° *Poetry and disappointment.*   In his boyhood Hamilton had translated Homer into blank verse, and, being gifted in rhyme and meter, had begun to compose poetry of his own. Hamilton's collected poetic efforts would fill a more-than-slim volume. The following poem, entitled "On College Ambition," is fairly representative of his attainments; it was written during his first year in college, when he was eighteen.

> Oh! Ambition hath its hour
> Of deep and spirit-stirring power;
> Not in the tented field alone,
> Nor peer-engirded court and throne;
> Nor the intrigues of busy life;
> But ardent Boyhood's generous strife,
> While yet the Enthusiast spirit turns
> Where'er the light of Glory burns,
> Thinks not how transient is the blaze,
> But longs to barter Life for Praise.

Look round the arena, and ye spy
Pallid cheek and faded eye;
Among the bands of rivals, few
Keep their native healthy hue;
Night and thought have stolen away
Their once elastic spirit's play.
A few short hours and all is o'er,
Some shall win one triumph more;
Some from the place of contest go
Again defeated, sad and slow.

What shall reward the conqueror then
For all his toil, for all his pain,
For every midnight throb that stole
So often o'er his fevered soul?
Is it the applaudings loud
Or wond'ring gazes of the crowd;
Disappointed envy's shame,
Or hollow voice of fickle Fame?
These may extort the sudden smile,
But they leave no joy behind,
Breathe no pure transport o'er the mind,
Nor will the thought of selfish gladness
Expand the brow of secret sadness.

Yet if Ambition hath its hour
Of deep and spirit-stirring power,
Some bright rewards are all its own,
And bless its votaries alone:
The anxious friend's approving eye;
The generous rival's sympathy;
And that best and sweetest prize
Given by silent Beauty's eyes!
These are transports true and strong,
Deeply felt, remembered long:
Time and sorrow passing o'er
Endear their memory but the more.

The "silent Beauty" mentioned in the last stanza was not just a poetical abstraction, but was a very real young lady whom Hamilton met through her brothers, who were fellow students at Trinity College. His feelings for the young lady led to an effusion of poetry. The lady, however, preferred another, and Hamilton suffered deep pangs of disappointment in his first love. One day, walking from the college to the observatory, he was strongly tempted to end everything in the still waters of the Royal Canal.

[More mathematicians than one might believe have been, or are, poets of one degree or another. As just one example, how many students of elementary geometry realize that Lorenzo Mascheroni (of compass-construction fame) is never considered in the encyclopedias as a mathematician, but as an accomplished Italian poet? An interesting paper could be written on mathematicians who were poets, with samples of their verses. One thinks of an essay or a brief anthology by, say, Clifton Fadiman.]

135° *Hamilton and Wordsworth.*   1827 Hamilton was appointed to the chair of astronomy at the University of Dublin when he was not yet twenty-two years old. Before assuming his professorial duties, he decided to tour England and Scotland. It was in this way that he first met the great poet William Wordsworth at the latter's home at Rydal Mount in Cumberland. The two men had a memorable midnight walk, oscillating between Rydal and Ambleside, and found it very difficult to separate. Each acquired a lifelong respect and admiration for the other. Later in life Wordsworth claimed that Coleridge and Hamilton were the two most wonderful men he had ever met.

In October of 1827, Hamilton moved into the Dunsink Observatory to commence his new duties. One of his first distinguished visitors was Wordsworth, and in commemoration of the visit one of the shaded walks in the gardens of the observatory was named "Wordsworth Walk." Hamilton, with his deep desire to be a poet, must have pressed his friend for encouragement. But Wordsworth advised Hamilton to concentrate his remarkable powers on science, and not long after his visit wrote to Hamilton as follows:

> You send me showers of verses which I receive with much pleasure, as do we all; yet have we fears that this employment may seduce you from the path of science which you seem destined to tread with so

much honor to yourself and profit to others. Again and again I must repeat that the composition of verse is infinitely more of an art than men are prepared to believe, and absolute success in it depends upon innumerable *minutiae* which it grieves me you should stoop to acquire a knowledge of. . . . Again I do venture to submit to your consideration, whether the poetical parts of your nature would not find a field more favorable to their exercise in the regions of prose; not because those are humbler, but because they may be gratefully and profitably trod, with footsteps less careful and in measures less elaborate.

The truth of the matter is that Hamilton, though he possessed poetic imagination, lacked poetic technique. Of course, *imagination* is one of the ingredients strongly found in poetry and in mathematics. We recall the statement made by Weierstrass: "It is true that a mathematician, who is not somewhat of a poet, will never be a perfect mathematician."

136° *A rare compliment.* When, in 1845, Hamilton attended the second Cambridge meeting of the British Association, he was lodged for a week in the sacred rooms of Trinity College in which tradition asserts that Isaac Newton composed his *Principia.* Hamilton was greatly affected by this compliment, and when he returned to Ireland he was fired with enthusiasm to try to prepare a work on quaternions that would not unworthily compare with Newton's *Principia.*

137° *Hamilton's second love affair.* Among Hamilton's pupils was the young Lord Adare, the eldest son of the Earl of Dunraven. While visiting Adare Manor one time, Hamilton was introduced to the De Vere family, who lived close by at Curragh Chase. The attractive daughter, Ellen de Vere, awoke amoral feelings in Hamilton, and he was soon composing sonnets and love poems to the young lady. His suit was encouraged by the Countess of Dunraven and favorably received by both of the young lady's parents. Things reached the point when he was about to propose marriage when the young lady made an incidental remark that restrained him and brought the affair to an end; she commented that she "could not live happily anywhere except at Curragh."

138° *A bad choice, and alcohol.*   Hamilton failed twice in love, and then made an unfortunate third choice. The lady this time was a Miss Bayly, who frequently visited her sister near the Dunsink Observatory where Hamilton lived. She was weak in mind, spirit, and body, highly inefficient, a poor manager, and a hypochondriac. The wedding took place, and Hamilton, who needed a wife of exactly the opposite qualities, had to spend the rest of his life enduring irregular and often forgotten meals, excessive and mislaid bills, and unbelievably attrocious and inefficient housekeeping.

By nature a convivial man, Hamilton took more and more to the solace and stimulation of alcohol, gradually expanding his intemperance from the privacy of his study to the open glare of the banquet hall. At one time, in his late thirties, he completely lost control of himself at a dinner of a scientific society in Dublin. When he sobered up, he was mortified at his public performance and he resolved thenceforth to abstain totally from all alcohol. He kept his resolution for two years, and then, at another scientific party, given by Lord Rosse, the Greenwich astronomer Airy taunted him for being a teetotaler. He broke his resolution, and from that time on gave in readily to his craving for alcohol.

When Hamilton died, at the age of sixty, his study-room floor was found heaped with piles of manuscripts, with little narrow paths between the piles. When the piles were carried out for examination, enough plates and dishes for a large family were found sandwiched between the sheets, bearing crusted and rotted remains of food. Unappetizing and poorly prepared meals had been thrust in upon Hamilton and nauseatingly left untasted, and his study room had developed into a pigsty.

E. T. Bell subtitled his excellent biographical chapter on Hamilton, "An Irish tragedy."

139° *A rare honor.*   Hamilton died, after several months of suffering, on September 2, 1865, at the start of his sixty-first year. During this final illness he received a singular honor from the United States. While the Civil War was in progress, the National Academy of Sciences was founded, and Hamilton received the news that he had been elected one of the ten foreign members and that his name had been voted to be placed the first on the list. Thus Sir William Rowan Hamilton was the first foreign associate of the National Academy of Sciences of the United States.

140° *Hamilton's wish.* "I have very long admired Ptolemy's description of his great astronomical master, Hipparchus, as a labor-loving and truth-loving man. Be such my epitaph."

141° *Modified Irish blarney.* H. J. S. Smith was a brilliant talker, a wit, and always in social demand. Alexander Macfarlane has reported some instances of Smith's wit. When someone mentioned the enigmatical motto of Marischal College, Aberdeen, "They say, what say they; let them say," Smith reacted, "Ah, it expresses the three stages of an undergraduate's career. 'They say'—in his first year he accepts everything he is told as if it were inspired. 'What say they'—in his second year he is skeptical and asks questions. 'Let them say' expresses the attitude of contempt characteristic of his third year." Smith once greeted a friend, "You take tea in the morning; if I did that I should be awake all day." Of a writer he said, "He is never right and never wrong; he is never to the point." Of the astronomer Lockyer, who served for many years as the editor of the scientific journal *Nature*, he said, "Lockyer sometimes forgets that he is only the editor, not the author, of *Nature*." Giving advice to a newly elected fellow of his college, he said, "Write a little and save a little; I have done neither." In 1881, at the jubilee meeting of the British Association held at York, Professor Huxley strolled down to the fine cathedral known as the Minster and met Smith coming out. Smith put on an expression of mock surprise. "You seem surprised to see me here," said Huxley. "Yes," replied Smith, "going in, that is; I would not have been surprised to see you on one of the pinnacles." Macfarlane said that socially Smith "was an embodiment of Irish blarney modified by Oxford dignity."

142° *On the turn of a penny.* H. J. S. Smith in 1849 carried off the highest honors in both the classics and mathematics at Oxford. It is said that he decided between these two areas of study for the field of his life's work by tossing a penny.

Smith advocated an antiutilitarian view of mathematics and once, at a banquet of the Red Lions, proposed the toast, "Pure mathematics, may it never be of any use to anyone." It has been reported that in one of his lectures, after he had furnished a new solution to an old problem, he commented, "It is the peculiar beauty of this method, gentlemen, and one which endears it to the really scientific mind, that under no circumstance can it be of the smallest possible utility." It must be admitted, however, that Smith made such exaggerated statements as a kind of defiance to the grossly utilitarian view of mathematics prevalent in England at the time. He also gave, on more than one important occasion, noted talks on the valuable and intimate union between mathematics and physics.

143° *His last act.*   H. J. S. Smith was a consistent liberal in politics, in university administration, and in religion. When, in 1882–1883, political agitation had grown favoring the extension of the franchise in the county constituencies, Smith supported the movement. Though he was suffering from a severe cold, at a meeting in the Oxford Town Hall he delivered a ringing speech on behalf of the movement, urging justice for all classes. He left the speaker's platform and made his way home to die of complications induced by the exposure and the excitement. He passed away on February 9, 1883, while in his fifty-seventh year.

TWO SCOTSMEN

MUCH of the mathematics of Scotland in the nineteenth century emanated from the University of Edinburgh, supplied by the minds and pens of some very able men. Among these were Peter Guthrie Tait (1831–1901), mathematical-physicist and ardent advocate of Hamilton's quaternions, and George Chrystal (1851–1911), master algebraist. The second of these two men was mentioned in passing in Item 333° of *In Mathematical Circles*, and the first one figures in Items 119°, 277°, 346°, and 349° of *Mathematical Circles Revisited*. Here are fourteen more tales about these two colorful professors. Some of these tales have been preserved for us by J. M. Barrie in his delightful and little-known *An Edinburgh Eleven, Pencil Portraits from College Life*. Though this is a very early work of Barrie's, one finds in it much of the charm, humor, sentiment, and whimsy that later characterized his famous novels and plays.

144° *Wiping away a stigma.* J. M. Barrie has told a few interesting stories about Peter Guthrie Tait, whom he had as Professor of Natural Philosophy when he was a student at the University of Edinburgh. One story is to the effect that when Tait himself was a student at Cambridge University, the mathematicians there were taunted for never excelling in Scriptural knowledge. Tait and a fellow mathematics student resolved to remove this stigma, and for two consecutive years each of them in turn captured the first prize in Biblical knowledge.

145° *T and T.* *The Elements of Natural Philosophy,* by William Thomson and P. G. Tait, was a famous text and came to be referred to as *T and T* (sometimes as *T and T'*). Barrie says that in his time at the University of Edinburgh, *T and T* was better known as *The Student's First Glimpse of Hades.* But Barrie ranked Tait not only as the most superb of demonstrators, but also as beyond doubt the finest lecturer at the University.

146° *Tait and Stevenson.* When Robert Louis Stevenson's biographical memoir of Fleeming Jenkin appeared, Tait criticized it at length, claiming that a sketch of a scientific man should be done by a scientific man. To this view Barrie has remarked that though scientific men may be the only ones who have something to say, they are also the only ones who can't say it.

147° *Holding back the crowd.* When students crowded too near a demonstration experiment conducted by Tait, he would squirt water on them from a tube.

148° *Keeping the demonstration desk clear.* There is a legend about the natural philosophy classroom at the University of Edinburgh that may or may not antedate Tait. It seems that the professor became annoyed by the students' habit of leaving their hats on the end of his demonstration desk when they entered the room. He accordingly warned them that the next hat he found there he would cut into pieces before their very eyes. The warning took effect, until one day when the professor had to leave the room for a few moments to get a piece of apparatus. During his absence one of the students quickly slipped into the closet where the professor hung his own hat and coat, secured the hat, and put it on the end of the demonstration desk, and with a certain degree of panic attained his customary seat in the classroom just seconds before the professor returned. When the professor saw the hat, he paused and said, "Gentlemen, I warned you what would happen if ever again I found a hat on the end of my demonstration desk." Deliberately taking a penknife from his pocket, he proceeded to slit the hat into several pieces, which he then tossed into the sink, whereupon the students gave forth with a great cheer.

This story recalls old "Reddy" Echols, Professor of Mathematics at the University of Virginia in the early nineteen hundreds. He kept one window of his classroom open so that he could spit out the window every so often. Once, when the professor was momentarily called from the room, the students closed his window—with very messy results after the professor returned to the classroom.

149° *Behind the times.*   Tait was very conservative and constantly battled university reform. Of his students he said that the less they knew of his subject when they started his course, the less, probably, they would have to unlearn—a remark heard today from some college instructors of calculus in regard to the teaching of calculus in the high schools.

150° *The Tait's compass.*   The telling of some Tait stories to Professor Lee Swinford, the beloved punster of the Mathematics Department of the University of Maine, brought forth from him the following additional Tait story. Though not many people seem to be aware of it now, Professor Swinford asserted, Tait once invented a remarkably good and different kind of direction compass, which became very popular and was known as a Tait's compass. The compass was widely adopted by mariners, hunters, surveyors, armies, and others. But one day, due to some strange and never-accounted-for astronomical incident—perhaps an overabundance of sun spots—all the Tait's compasses in the world simultaneously went berserk. Mariners got off their courses, hunters became lost, surveyors committed serious errors, whole armies went astray. All of which gave rise to the familiar saying: "He who has a Tait's, is lost."

151° *The hare and the hounds.* J. M. Barrie recalled a gloomy student who sat before him in Professor George Chrystal's first-year mathematics class. The fellow hacked, "All hope abandon, ye who enter here," into his desk with his penknife. "It took him a session," said Barrie, "and he was digging his own grave, for he never got through; but it was something to hold by, something he felt sure of. All else was spiders' webs in chalk." Chrystal's lectures were apparently only for the gifted few and soon pulled away from the average student. He was, Barrie concluded, "a fine hare for the hounds who could keep up with him." The students "doggedly took notes, their faces lengthening daily. Their notebooks reproduced exactly the hieroglyphics of the blackboard, and, examined at night, were as suggestive as the photographs of persons one has never seen." Chrystal's classes were quiet, "nothing heard but the patter of pencils, rats scraping for grain, of which there was abundance, but not one digestion in a bench."

152° *An embarrassment.* Barrie turned in a paper of answers to the first weekly set of exercises in Chrystal's beginning mathematics class at the University of Edinburgh. When the papers were returned, Barrie was there "to accept fame—if so it was to be—with modesty; and if it was to be humiliation, still to smile." Professor Chrystal remarked that there was one paper in the set whose owner's name he could not read, and he sent the paper about the class to be deciphered. When the paper reached Barrie, he recognized it as his, but embarrassed by the miserable grade on it he pleasantly passed it along and it finally returned unclaimed to the professor.

153° *An allegory.* Barrie tells of an elderly man who for years daily read *The Times* from beginning to end. The man fell ill for a fortnight and got behind in his reading. Upon recovery he began with the copy of *The Times* where he had left off just before his illness. He struggled valiantly to catch up on *The Times*, but in vain. "This is an allegory for the way the students panted after Chrystal," says Barrie.

154° *Unpopular*.   Chrystal's lectures were so difficult for the ordinary student to follow that his classes were not greatly sought after. There is a legend that says one day on Chrystal's classroom door appeared the notice: "There will be no class today, as the student is unwell."

155° *Taking roll*.   Chrystal discovered that students can sit too near the classroom door. He accordingly took to calling roll in the middle of the hour to insure a continued attendance.

156° *Poor arithmetician*.   "In the middle of some brilliant reasoning," recalls Barrie, "Chrystal would stop to add 4, 7, and 11. Addition of this kind was the only thing he could not do, and he looked to the class for help—'20,' they shouted, '24,' '17,' while he thought it over. These appeals to their intelligence made them beam."

157° *The marble*.   The benches in Professor Chrystal's classroom rose in steps above the front floor. One day a student at the end of bench ten dropped a marble while the professor, with his back to the class, was working at the blackboard. The marble rolled downward toward the professor, very audibly falling down each of the ten steps on its way to the floor. Professor Chrystal never turned his head, but when the marble reached the floor, he said, still without turning his head, "Will the student at the end of bench ten, who dropped that marble, stand up?" While continuing his work at the blackboard, he had kept count of the falls of the marble from step to step.

THE LAST UNIVERSALIST

HENRI Poincaré (1854–1912) is generally regarded as the last of the universalists in the field of mathematics—that is, as the last mathematician of whom it can in a reasonable sense be claimed that *all* of mathematics was his province. Mathematics has grown at such an incredible rate in modern times that it is believed quite impossible for anyone ever again to achieve such a distinction. But this same view had been held before Poincaré began his mathematical career, when it was regarded that Gauss would surely have to be listed as the last of the universalists in the field of mathematics.

Here are some stories about Poincaré, who somehow or other received no mention in either of our two earlier collections.

158° *Poor motor coordination.* From infancy Poincaré was physically awkward and was never able to participate successfully in children's sports. When he learned to write it was discovered that he was ambidextrous; he could perform equally badly with either hand. It has been said that his later work in science might have been closer to reality if he had not been handicapped in the laboratory by lack of dexterity with his hands. In spite of his physical awkwardness, however, he did considerable hiking and climbing (because of his interest in geology), and he became an indefatigable dancer.

159° *The artist.* Poincaré had no ability whatever in drawing, and he earned a flat score of *zero* in this subject. At the end of the school year his classmates jokingly organized a public exhibition of his artistic masterpieces. They carefully labelled each item in Greek—"This is a house," "This is a horse," and so on. Because of his inability to draw, Poincaré lost first place in geometry, though he did secure second place.

160° *Fame helps.* In 1871, when he was seventeen, Poincaré took and passed the examinations for the degrees of bachelor of letters and bachelor of science. But he almost failed in mathematics! He had arrived late for the examination and became flustered, with the result that he bungled in giving the simple proof of the formula for the sum of a convergent geometrical progression. Fortunately the fame he had already earned in mathematics had preceded him. The chief examiner declared that any student other than Poincaré would have been rejected.

161° *How does he do it?* Since Poincaré never took notes in class, his classmates thought he was a trifler who could bandy terminology but had no real grasp of the subject matter. To test him out, an

95

advanced student was one day delegated to quiz him on a particularly tough problem in mathematics. With no preliminary thought, Poincaré immediately furnished the solution. Later he again surprised his classmates by easily capturing, without any preliminary preparation, the first prize in mathematics in the entrance examinations for the School of Forestry. There is a famous story that when he later entered the École Polytechnique, the examiner, having been warned that young Poincaré was a mathematical genius, deferred the examination for almost an hour in order to devise some especially hard questions for this candidate. Poincaré effortlessly disposed of each question, won the highest grade, and passed in first place into the École Polytechnique. His off-the-cuff speed in solving mathematical difficulties led colleagues throughout his life to exclaim, "How does he do it?"

162° *A misunderstanding*.   Henri Poincaré's paternal grandfather had two sons, Léon and Antoine. Léon became a first-rate physician and a professor of medicine and Antoine became inspector-general of the department of roads and bridges. Léon's son Henri became the world's leading mathematician of his time. One of Antoine's two sons, Raymond, went into law and became the President of the French Republic during World War I; Antoine's other son became director of secondary education.

A story is told that during the days of the War someone asked Bertrand Russell who he regarded as the greatest man produced by France in modern times. Without hesitation, Russell replied, "Poincaré." "What! *That* man?" exclaimed his astonished questioner in surprise, believing that Russell meant Raymond Poincaré, the President of the French Republic.

163° *Cousins*.   There was little love lost between Henri Poincaré and his cousin Raymond, at one time President of France. Once, when Henri was introduced to a group, someone asked, "Aren't you Raymond Poincaré's cousin?" "Not at all," answered Henri, "he is *my* cousin." [See Item 344° of *In Mathematical Circles* for a similar story about C. G. J. Jacobi and his brother M. H. Jacobi.]—MAXEY BROOKE

164° *Literary success*.   Poincaré was not only one of the foremost creators of mathematics and science, he was also one of the ablest popularizers of these subjects. His inexpensive paperback expositions were avidly bought and widely read by people of all walks of life. They are masterpieces that, for lucidity of communication and engaging style, have never been excelled, and they have been translated into many foreign languages. So great was the literary excellence of Poincaré's popular writing that he was awarded the highest honor that can be conferred on a French writer—he was elected a member of the literary section of the French *Institut*, an honor truly envied by any aspiring French poet or novelist.

165° *A poor administrator.*   Many an eminent mathematician has shown himself also to be an expert administrator, and frequently a top-flight mathematician at a university ultimately becomes a dean or the president of his institution. Poincaré seems completely to have lacked this type of ability, a skill that his cousin Raymond, who became President of the French Republic, possessed in such great abundance.

166° *A powerful and unusual memory.*   Poincaré could read at incredible speed, and whatever he read became a permanent possession of his mind. In retention and recall he even exceeded the fabulous Euler (see Item 242° of *In Mathematical Circles*). He was particularly strong in spatial memory, and could always recall the very page and line of a book where any particular statement had been made. Another of his peculiarities, brought on by his poor eyesight, was his ability to absorb theorems and passages of mathematics chiefly by ear, rather than eye, as do most other mathematicians. He developed this ability in school, where, unable to see the blackboard well, he sat back and listened, following and remembering everything without taking any notes.

167° *Poincaré's manner of working and composing mathematics.* Poincaré worked his mathematics in his head while restlessly pacing about, and when it was completely thought through, he committed it to paper in a dash and with essentially no rewriting or erasures; noise did not disturb him while he worked. Cayley, and perhaps Euler, too, composed mathematics in the same way. Marks of this hasty composition show in some of Poincaré's papers. In contrast one recalls the meticulously prepared productions of Gauss, and Gauss's motto: "Few, but ripe."

168° *Rated an imbecile.* When Poincaré was at the height of his powers, both as a creator and as a popularizer of mathematics and science, he submitted to a battery of Binet intelligence tests. His showing on these tests was so disgraceful that he was rated an imbecile.

It might be that most intelligence tests will fail properly to judge a genius, for a genius frequently sees more deeply into the questions of a test than does the originator or administrator of the test. For example, a common question that used to appear on the Binet tests (before the shortcomings of the question were pointed out) ran somewhat as follows. "It is now twenty minutes past eight. What time would the clock read if the two hands of the clock should be interchanged, the big hand put where the little hand is and the little hand put where the big one is?" The expected answer is, "Twenty minutes to four," but the correct answer is, "No time at all." For if the hands are interchanged, the little hand would be exactly on the four. But if the little hand is exactly on the four, the clock must read precisely four o'clock, and the big hand would have to be on the twelve, and not down near the eight.

169° *Two kinds of absentmindedness.* Poincaré was absent-minded in two different ways. One way was genuine, such as the time when he (really) quite unintentionally packed and carried off some hotel linen in his traveling bag. The other way was for convenience, such as the time he left a distinguished, but unwanted, visitor waiting for three hours behind some drapes that divided his study, although the visitor knew Poincaré had been notified of his presence. The visitor waited patiently without making a sound. Finally, in hopes of getting better results, Poincaré thrust his head through the drapes, bellowed, "You are greatly disturbing me," and then quickly withdrew behind the drapes again. Upon this, the visitor got up and departed, leaving the "absentminded" professor to his work.

Poincaré frequently forgot his meals and hardly ever could remember if he had or had not breakfasted. Since he was not overly

fond of eating, as eating too often interfered with more important
things, it is hard to say which of his two kinds of absentmindedness
this was.

170° *A lover of animals.*   Like William Rowan Hamilton (see
Item 337° of *In Mathematical Circles*), Poincaré was a genuine lover of
animals. The first time he experimented with a rifle he accidentally
killed a bird at which he was not intentionally aiming. He was so
deeply affected by this mishap that he refused (except with great
distaste in compulsory military drill) ever again to handle a firearm.

171° *An act of heroism.*   Because of his interest in geology,
Poincaré at one time considered becoming a mining engineer, and
accordingly entered the School of Mines after he graduated from the
Polytechnique. During his apprenticeship, an explosion and fire
occurred in the mine he was connected with, and sixteen lives were
taken. Immediately after the explosion, Poincaré descended into the
mine as part of a rescue crew.

172° *Sylvester meets Poincaré.*   In Item 143° of *In Mathematical
Circles*, we described the first meeting between Briggs and Napier, as
recorded for us by the astrologer William Lilly in the *History of his Life
and Times*. It will be recalled that upon meeting, the two mathematicians stared with admiration at each other for almost a quarter of an
hour before speaking. J. J. Sylvester alludes to this historic meeting
when describing the first time he met the famous Henri Poincaré.

"I quite entered into Briggs' feelings at his interview with Napier
when I recently paid a visit to Poincaré in his airy perch in the Rue
Gay-Lussac. . . . In the presence of that mighty reservoir of pent-up
intellectual force my tongue at first refused its office, and it was not
until I had taken some time (it may be two or three minutes) to peruse
and absorb as it were the idea of his external youthful lineaments that
I found myself in a condition to speak."

173° *An unfinished symphony.*   In 1911 Poincaré had a premoni-
tion that he would not live much longer. On December 9th he wrote a
letter to the editor of a mathematics journal asking if, contrary to

99

custom, it would be possible to have accepted for publication an unfinished paper on a problem that he regarded as of great importance. He explained to the editor that at his age he might not be able to complete the solution of the problem, but he felt his partial results might put some researcher on the right path. The paper was accepted and published, and, not long after, the solution was completed by the twenty-seven year old American mathematician George David Birkhoff (1884–1944). Poincaré died suddenly on July 17, 1912.

CROUTONS FOR THE FRENCH SOUP

WE complete the present quadrant with a small handful of stories about some French mathematicians, amateur and professional. Napoleon Bonaparte (1769–1821) has already been considered in several Items of *In Mathematical Circles*, and Camille Jordan (1838–1922) and the polycephalic Bourbaki in a number of Items of *Mathematical Circles Revisited*.

174° *Deep insight.* During the Egyptian campaign, Napoleon occasionally sat down to chat with the academicians on the expedition. Once, in an expansive mood, he confessed that he had missed his real vocation in life. He had wanted to be a scientist and emulate Newton. "But, mon général," said Monge, "that is impossible. Newton has already found the principle of the universe." "No," said Napoleon, "Newton has only found what governs the world in the large. What still has to be found is the principle that governs the world in the small—the laws of the atom."—DIRK J. STRUIK

175° *Precision.* Napoleon certainly had a precise mathematical way of expressing himself. My favorite example is the article of the Code Napoléon that reads, "L'enfant concu pendant le marriage a pour père le mari."—DIRK J. STRUIK

176° *Jordan and notation.* Camille Jordan employed a certain complexity of notation in his mathematical writing, and it has been said that if he had four quantities all on the same standing (such as a, b, c, d or α, β, γ, δ) they would appear as a, $M_3{}'$, ε_2, $\Pi_{1,2}{}''$.

177° *A story of Hadamard's youth.* Jacques Hadamard (1865–1963), because of his kinship with the wife of Colonel Dreyfus of the famous Dreyfus affair in France, was an ardent Dreyfusard. On the other hand, the great mathematician Charles Hermite was a conservative and an equally ardent anti-Dreyfusard. Now it happened that Hermite was to examine young Hadamard for the doctorate. Because of their political differences, Hadamard felt considerable nervousness as the time for the examination approached, and his nervousness certainly was not relieved when at the start of the official occasion Hermite growled, "M. Hadamard, you are a traitor!" As Hadamard began to mumble something in confusion, Hermite continued, "You have deserted geometry for analysis."

178° *A story of Hadamard's older age.* A notice of the passing of Jacques Hadamard in October 1963, at the very old age of ninety-eight, appeared in the *Bulletin de la Société de Mathématique de France.* The notice was accompanied by a fine portrait of Hadamard as he looked to his friends for so many, many years. The sparse and almost stringy beard, the hooked nose, the scarce and closely clipped hair, and the thinnish, friendly, sensitive features recalled a story told by Norbert Wiener. When Professor Wiener was lecturing in China in 1935, his friend Hadamard, then seventy, arrived for a prolonged stay at the beginning of the second semester. One day, while rummaging in the antique shops of Peiping, Wiener came across a Chinese-ancestor portrait that remarkably resembled Hadamard—so much so that anyone familiar with Hadamard's features would immediately notice the very close resemblance. Wiener bought the portrait and gave it to Hadamard. Though Hadamard appreciated the picture very much, it somehow later became lost or misplaced, and on subsequent visits by Wiener to the Hadamards the picture could not be located. Professor Wiener felt that Mme. Hadamard did not admire her husband in the role of the conventional nodding mandarin, and so secretly disposed of the likeness.

179° *Bourbaki's office.* One of the oldest buildings on the Champaign-Urbana campus of the University of Illinois is Altgeld Hall. Originally the library, this building later housed the Law School and the Mathematics Department, and most recently has been taken over completely by the Mathematics Department. Over the years renovations, including additional partitions and floors, have taken place. However, not all of these have come out evenly. In particular, there is one small room, without a window, located between floors, access to which is possible only by an emergency stop on the elevator.

Shortly before a regional meeting of the American Mathematical Society on the campus, a university official was inspecting the building before the visiting mathematicians arrived. To his horror, he discovered the small cobwebbed and dusty room had been furnished with a broken-down desk and rickety chair, an empty wine bottle holding a candle dripping wax, an antique inkwell, and a long quill pen. The small sign attached to the entrance read: N. Bourbaki.

JOHN W. TOOLE

180° *Mathematicians are like Frenchmen.* Goethe, in *Maximen und Relexionen*, said, "Mathematicians are like Frenchmen: whatever you say to them they translate into their own language and forthwith it is something entirely different."

QUADRANT THREE

*From a misplaced masterpiece
to a comparison of two deaths*

TWO NORWEGIANS AND A RUSSIAN

NIELS Abel (1802–1829), who died in his twenty-seventh year, is regarded as Norway's greatest mathematician; we start this new quadrant by telling three more stories about him. Marius Sophus Lie (1842–1899), of transformation-group fame, was another eminent Norwegian mathematician, and we tell an anecdote about him. We then conclude this short introductory section with three stories about the Russian mathematician Nicoli Ivanovich Lobachevsky (1793–1856), the creator of the first non-Euclidean geometry.

181° *A lost manuscript.* Few things can be more exasperating to an author than to have his manuscript lost or misplaced after he has submitted it for possible publication. Niels Abel once left one of his greatest masterpieces, *Memoir on a general property of a very extensive class of transcendental functions*, in Cauchy's care to be presented to the French Academy for judgment and possible publication. Cauchy, being very busy with researches of his own, passed the paper over to Hachette, who made the presentation to the Academy on October 10, 1826. The outcome was that Legendre and Cauchy were appointed to referee the paper—Legendre was seventy-four and Cauchy was thirty-nine. A couple of years then passed, during which Abel heard nothing about his paper. He finally wrote to Jacobi, telling of the paper and what was in it. On the fourteenth of March, 1829, Jacobi wrote to Legendre in an effort to find out the fate of the remarkable paper. Legendre replied that the manuscript was barely legible, the ink hardly visible, and the letters badly formed, and that it was agreed the author should be asked to submit a more readable copy. Cauchy took the paper home, presumably to communicate with Abel on its rewriting, misplaced the paper and soon forgot all about it. In time the author's unheeded enquiries about his paper became known in Norway, and the Norwegian consul at Paris, in an effort to stir up the Academy and secure some sort of report on the disposition of the paper, raised a minor diplomatic fuss about the missing manuscript. Upon this, Cauchy managed to resurrect the paper, and in 1830 the French Academy made amends for all its bungling and awarded Abel, jointly with Jacobi, the Grand Prize in Mathematics. But Abel was then dead.

It was not until 1841 that Abel's epoch-making paper was finally published. Even here some inexcusable bungling took place. For, somehow or other, before the proof-sheets could be read, the editor and the printer managed between them to lose the original manuscript!

182° *By studying the masters.*   Once when asked his formula for so rapidly forging ahead to the first ranks of his discipline, Abel replied, "By studying the masters and not their pupils."

183° *Spurred on because of a bully.*   Isaac Newton, it will be recalled (see Item 192° of *In Mathematical Circles*) received his first stimulus for serious school study when he was kicked in the stomach by a bully and fellow schoolmate. After roundly defeating the bully in a fight, Isaac was determined also to beat him at the books.

Niels Abel, when a youngster in school, was stimulated to serious study, not through brutality to his own person, but to that of a fellow student. The student was so unmercifully flogged by the sadistic schoolmaster that he died. The school board found itself forced to remove the brutal teacher from his position. He was replaced by Bernt Michael Holmboë (1798–1850), a fine pedagogue and a competent mathematician. Holmboë soon recognized deep mathematical talent in the young Niels and inspired the lad to hard work and accomplishment. Holmboë became Abel's lifelong and devoted friend, and he did all he could to assist the young man financially and professionally. It was Holmboë who later in 1839 edited the first edition of Abel's collected works.

184° *An unfortunate incident.*   The Norwegian mathematician Marius Sophus Lie and the Prussian mathematician Felix Klein first met in Berlin in the winter of 1869–70, and the two mathematicians there published some joint papers. In the summer of 1870, Lie and Klein were together in Paris and in close touch with the two French mathematicians Camille Jordan and Jean Gaston Darboux. This fruitful gathering of mathematicians of different nationalities was brought to a sudden end by the outbreak of the Franco-Prussian War.

Klein immediately left Paris for Germany, and Lie started out afoot to walk through France to Italy. Looking suspicious, Lie was arrested as a spy and held in prison for a month until Darboux was able to obtain his release.

185° *An almost unique event.* When the Russian government decided to modernize the old buildings on the Kazan University campus and to construct some new ones, Lobachevsky became very concerned that the work be done well and with no wasting of the appropriation. To fit himself properly for the task, he learned architecture, mastering it so practically that not only were the new and the renovated buildings very attractive and highly functional, but they were constructed for less money than had been appropriated.

186° *A most admirable trait.* Lobachevsky had such a pride in, and love for, his university that even when he had attained the dignity of Rector of the institution, he would not hesitate to doff his coat and collar and do the most menial work if such labor would result in the improvement of the university. There is a story told that a distinguished foreign visitor, coming upon the coatless Rector scrubbing some floors, took the worker for a janitor and asked to be shown through the library and the museum of the university. The visitor was much impressed by the courtesy and high intelligence of the janitor, and on completion of the tour proffered a handsome tip. To the visitor's surprise, the janitor indignantly refused, and the visitor put the behavior down as one more eccentricity of the high-minded janitor. That evening he and Lobachevsky were introduced to one another at the Governor's table, and apologies then issued from both sides.

Would that all administrators, faculty, and students of a college take a similar pride in their institution. The result would be halls constantly clean of paper scraps and cigarette butts, and a campus free of beer cans, candy wrappers, broken lights, and mutilated shrubbery.

During two years spent at the Gorham campus of the University of Maine, the writer of these stories lived next door to the retired President, Francis L. Bailey, a fine and remarkable man who, though retired from the headship of the college, never failed to pick up any litter that he passed as he walked about the campus and the town, and in the fall he donned his old clothes and helped the grounds crew rake up the fallen leaves.

187° *An apt description.*   William Kingdon Clifford had a happy knack of coining apt descriptive phrases. Thus it was he who first called Lobachevsky (the creator of the first non-Euclidean geometry), "the Copernicus of geometry."

THE PRINCE OF MATHEMATICIANS

CARL Friedrich Gauss (1777–1855) was easily the most eminent mathematician of the nineteenth century and one of the very greatest mathematicians of all time. He gave to Göttingen University such an aura in the field of mathematics that that institution became and remained for years the Mecca and dream of students of the subject, and there developed at Göttingen one of the truly great schools of mathematics of modern times. Stories and anecdotes were bound to accumulate around such an influential founder and such a towering genius. We have already, in our two earlier collections, told over a dozen stories about Gauss; here are a couple of dozen more.

188° *A student's lamp.*   On cold winter nights Carl Friedrich Gauss and his older brother Georg were sent to bed early by their father in order to save light and heat. Carl would take a turnip with him up to his attic room. By hollowing out the turnip, rolling a wick of rough cotton for it, and using some fat for fuel, he obtained a dim light by which he studied far into the night until driven to bed by cold and fatigue.

189° *Lost fame.*   One day, in the yard of the palace, the Duchess of Brunswick came across young Carl Friedrich Gauss absorbed in a book, and she was much amazed to find that the little boy fully understood what he was reading. Upon receiving a report of the incident, the Duke sent a messenger to the Gauss home to bring the boy to the palace. When the messenger arrived at the Gauss home, it was mistakenly thought that the invitation was for Carl's older brother

Georg, who resisted with much crying and carrying on. Then it was discovered that the invitation was not for Georg, but for Carl. As a result of Carl's visit to the palace, the impressed and kindly Duke of Brunswick furnished the financial means for furthering the boy's education. After Carl became world-famous and Georg was but a common laborer, Georg is alleged to have said, "Yes, if only I had known, then I would be a professor now; the invitation was offered to me first, but I didn't want to go to the castle."

190° *Ribbentrop as an overnight guest.*   Among Gauss's friends was Georg Julius Ribbentrop (1798–1874), a professor of law at Göttingen University. Ribbentrop was a confirmed bachelor, a campus eccentric, and the typical absentminded professor. Some amusing anecdotes have been told about Ribbentrop in connection with Gauss. Thus there was the time Ribbentrop was invited to Gauss's home at the Göttingen observatory for an evening dinner. After the dinner was over, a long and heavy thunderstorm came up. Since the observatory was a considerable distance from Ribbentrop's rooms in town, he was invited to stay the night at his host's home. Ribbentrop accepted the invitation, but soon disappeared when no one was noticing. Some time later the observatory doorbell rang, and there in the rain before his astonished host stood Ribbentrop, soaked to the skin. He had hurried home to fetch his sleeping equipment for the overnight invitation.

191° *Ribbentrop as an observatory guest.*   A total eclipse of the moon occurred on the evening of November 24, 1836, and Gauss had promised to show it, through the observatory telescope, to his friend Ribbentrop. On the appointed evening a pouring rain had set in, making the observation impossible. Gauss assumed that his guest would not appear, and so was considerably surprised when, suddenly, Ribbentrop, thoroughly drenched, stood at his door.

"My dear colleague," Gauss said, "in this rain our planned observation of the eclipse will be impossible."

"Not at all," replied friend Ribbentrop, flourishing a large umbrella, "My landlady saw to it that this time I did not forget my umbrella."

192° *Gauss's assistant.* The optician J. H. Teipel served as Gauss's assistant at the Göttingen observatory. Among his duties was that of showing visitors through the observatory, giving them a view through the telescope, and satisfying them with popular explanations of the heavenly bodies. One evening, while showing a group some of the planets, a lady inquired as to the distance between the earth and Venus. "I cannot tell you that, Madame," explained Teipel. "Hofrat Gauss takes care of the *numbers* in the heavens; I call attention to the *beauties* in the sky."

193° *Gauss's power of concentration.* Carpenter, in his *Mental Physiology*, tells a story about Gauss's power of concentration that is probably apocryphal, for Gauss has never been described as the absentminded-professor type. According to the story, Gauss was engaged in some profound investigation when a servant told him that his wife, for whom he had a deep attachment and who was seriously ill, had grown worse. Gauss seemed to hear, but he continued his work. After a while, the servant appeared again, announced that his mistress was much worse, and begged the professor to come at once. "I will come presently," Gauss replied, but again lapsed into his concentrated train of thought. A little later, the servant appeared a third time and stated that his mistress was dying and that if the professor did not come at once he might not find her alive. Gauss raised his head and calmly replied, "Tell her to wait until I come."

194° *Source of pleasure.* Gauss had six children, four sons and two daughters. The fourth child, Peter Samuel Marius Eugenius Gauss—always called Eugene—lived to be eighty-five and was the last survivor of the children. He was the most talented of all the children, showing some promise in both mathematics and languages. Once, when only a boy, Eugene told his father of the delight he experienced in solving a certain problem in grammar. The father's eyes brightened and he replied, "Yes, son, the pleasure one receives from the solution of such problems is very great, but it cannot be compared with the similar pleasure one derives from the solution of mathematical problems."

195° *Maintaining a standard.* After Gauss had felt out his sons' abilities, he did not want them to attempt specialization in mathematics, for he was convinced none of them would surpass him, and he did not wish the name of Gauss in the field of mathematics to be lowered.

196° *Some similarities between father and son.* Eugene Gauss, like his great father, had a penchant for languages. When only a boy he mastered French almost perfectly. Indeed, later in life, when in America, he spoke French so fluently that he was occasionally taken to be a Frenchman. When he was in the employ of the American Fur Company, at the headwaters of the Mississippi and Missouri rivers, he easily learned to speak the Sioux tongue and assisted a missionary by the name of Pond in constructing a Sioux alphabet and in translating the Bible into the Sioux language. In 1895, the year before he died, he spoke of his lifelong desire to study philology, and ruminated that had he stayed in Europe he might have become a professor of philology.

Again like his great father, Eugene Gauss found a certain enjoyment in performing long arithmetical calculations mentally. As an example, when he was over eighty years old and had become blind, he mentally computed the amount to which one dollar would grow over a period of 6000 years if compounded annually at the rate of 4 per cent. If the resulting sum should be put in the form of a cube of gold, the mass would be so enormous that it would take light some quadrillions of years to travel the length of the edge of the cube.

197° *Gauss and his grandson.* Gauss had a number of grandchildren but got to see only one of them, for the others lived in America. One day the German grandson was playing in the garden at the Göttingen observatory, where Gauss lived. Coming upon the child, the grandfather asked, "What do you expect to make of yourself?" The young boy replied, "Well, Grandfather, what do you expect to make of *yourself*?" The old man patted the boy on the shoulder and smilingly said, "My boy, I am already somebody."

198° *Gauss and the British Admiralty.* The following story is told by Augustus De Morgan in his *Budget of Paradoxes*.

Francis Baily (1774–1844), a London stockbroker with an interest in science and astronomy in particular, in 1835 wrote a book entitled

An Account of the Rev. John Flamsteed, the First Astronomer-Royal. The book was published for distribution by the British Admiralty, and Baily was entrusted with the task of drawing up the distribution list. Just before the book appeared, rumors as to its extraordinary revelations caused many persons of influence to apply to the Admiralty for copies. With an insufficient supply of copies, the Lords of the Admiralty found themselves in a difficulty, but on examining Baily's distribution list they found names of people they considered so obscure as not to merit receiving copies of the book. Mr. Baily was summoned and the matter put to him. Upon inquiring as to which names were found undeserving, the Admiralty secretary said, "Well, now, let us examine the list; let me see; now—now—come! Here's Gauss, who's Gauss?" Mr. Baily had to inform the secretary that Gauss was considered to be the greatest living mathematician.

199° *The Berlin Academy of Sciences.*   In 1805 Frederick William III of Prussia invited Alexander von Humboldt to membership in the Berlin Academy of Sciences, so that the Academy might shine from the luster of his presence. Humboldt replied that his presence in the Academy would be of no importance, and that the *only man* who could give new splendor to the Berlin Academy was Carl Friedrich Gauss. On April 18, 1810, Gauss was elected a member of the Berlin Academy of Sciences.

200° *Wrong order.* Adolf Friedrich, a Göttingen alumnus and Duke of Cambridge, once said to Alexander von Humboldt, "Göttingen is often criticized, but as long as we have the library and Gauss, we can ignore it." Humboldt replied, "I agree, but it is my duty to request your royal highness to change the order of rank of the treasures and to name first the foremost mathematician of our age, the great astronomer, the brilliant physicist."

201° *Theory and practice.*   H. B. Lübsen (1801–1864), a private teacher of mathematics in Hamburg, wrote highly successful elementary textbooks for self-instruction in algebra and arithmetic. He also wrote a textbook on higher geometry. This latter he dedicated to the Spirit of Descartes, and in the foreword to the first edition wrote,

"Theory, said my revered teacher Gauss, attracts practice as the magnet attracts iron."

202° *A sacrilege.* It was on July 16, 1799, that Gauss was awarded his doctorate from the University of Helmstedt, after offering his famous dissertation containing the first rigorous proof of the fundamental theorem of algebra. This material was incorporated, with other gems, in his great *Disquisitiones arithmeticae*, published September 29, 1801.

On July 16, 1849, exactly fifty years after the awarding of his doctorate, Gauss enjoyed the celebration at Göttingen of his golden jubilee. As a part of the "show," Gauss, at one point of the proceedings, was to light his pipe with a piece of the original manuscript of his *Disquisitiones arithmeticae*. Dirichlet, who was present at the celebration, was appalled at what seemed to him a sacrilege. He boldly rescued the paper from Gauss's hands and treasured the memento the rest of his life; it was found by his editors among his papers after he died.

203° *The hunt rather than the treasure.* In a letter dated September 2, 1808, to his friend Wolfgang Bolyai, Gauss wrote:

> It is not knowledge, but the act of learning, not possession but the act of getting there, which grants the greatest enjoyment. When I have clarified and exhausted a subject, then I turn away from it, in order to go into darkness again; the never-satisfied man is so strange— if he has completed a structure, then it is not in order to dwell in it peacefully, but in order to begin another. I imagine the world conqueror must feel thus, who, after one kingdom is scarcely conquered, stretches out his arms for others.

204° *Gauss and a student.* One day Gauss encountered a student staggering along a street of Göttingen. On approach the student endeavored to straighten up, but with only partial success. Gauss eyed him critically, pointed a finger at him, and smilingly said, "My young friend, I wish science would intoxicate you as much as does our good Göttingen beer!"

205° *The onerous side of teaching.* In a letter dated January 7, 1810, to his astronomer friend Friedrich Wilhelm Bessel, Gauss wrote:

This winter I am teaching two courses for three listeners, of whom one is only moderately prepared, one scarcely moderately prepared, and the third lacks preparation as well as ability. These are the *onera* of a mathematical profession.

206° *Cayley on Gauss.*   Arthur Cayley greatly admired Gauss:

All that Gauss has written is first rate; the interesting thing would be to show the influence of his different memoirs in bringing to their present condition the subjects to which they relate, but this is to write a History of Mathematics from the year 1800.

207° *Absolute and relative properties of a surface.*   A surface can be looked at in two ways: as the boundary of a solid body or as a detached two-dimensional film. The former is the way a construction engineer might regard the surface, and the latter the way a surveyor might regard it. The first viewpoint leads one to search out the properties of the surface that relate it to its surrounding space, and the second viewpoint leads one to search out the properties of the surface that are independent of its surrounding space. Properties of the first kind are called *relative properties* of the surface, and their study is called the *extrinsic geometry* of the surface; the properties of the second kind are called *absolute properties* of the surface, and their study is called the *intrinsic geometry* of the surface. It is interesting that the two great early contributors to the differential geometry of surfaces, Monge and Gauss, respectively saw a surface primarily as the boundary of a solid and primarily as a detached two-dimensional film. Monge is noted, among other things, for his work as a construction engineer of military fortifications, and Gauss is noted, among other things, for his work in geodesy and geodetic surveying.

208° *The heliotrope.* The heliotrope (or heliograph) was Gauss's favorite invention, and he claimed that he was led to it by meditation (in connection with his interest in geodetic surveying) and not by accident. From the steeple of St. Michael's in Lüneberg he once saw the sunlight reflected from a windowpane of a Hamburg steeple, but this, he maintained, was an after-incident that merely strengthened his conviction of the instrument's practicality. Gauss enjoyed telling how, when the heliotrope was first tested, a crowd of gathered spectators gave a shout of exultation when the distant reflected light first appeared.

209° *The Hohenhagen Tower.*   In 1818 Gauss was commissioned to make a geodetic survey of the kingdom of Hanover, a task that occupied him for several years and led to his invention of the heliotrope and his brilliant work on the differential geometry of surfaces. The actual triangulation of the survey covered the years 1821–23.

The most imposing memorial to Gauss is the Gauss Tower on Mt. Hohenhagen. It marks one vertex of the Brocken-Hohenhagen-Inselsberg triangle, so important in the Hanover survey. Built of basalt and topped with a red tile roof, the tower rises 120 feet above the ground and furnishes the visitor a splendid view of the surrounding country. In the tower there is a room containing some relics, some astronomical and surveying instruments, and a large marble bust of Gauss executed by Gustav Eberlein.

210° *The Gauss-Weber monument.*   It was in 1831 that Guass commenced collaboration with his colleague Wilhelm Weber in basic research in electricity and magnetism, and it was in 1833 that the two scientists devised their electromagnetic telegraph. This discovery is represented by a masterful monument created by Ferdinand Hartzer and erected in 1899 on the Göttingen campus. Built with a cylindrical base, it shows Gauss seated and Weber standing, the two men apparently discussing their telegraph. The monument was unveiled during elaborate ceremonies. The only error in the monument is the seeming closeness in age of the two men; actually, Gauss was twenty-seven years older than Weber.

211° *Ravages of war.*   To protect the great Göttingen library (one of the richest collections in Germany) from possible destruction by bombs during the Second World War, the volumes and incunabula were removed to what was considered a safe place. Göttingen emerged from the war unscathed, but a bomb accidentally found and destroyed the removed library.

The house in Brunswick in which Gauss was born, and which served as a museum, was destroyed in an air raid on October 15, 1944.

Fortunately, the contents of the museum had been removed and are now found in the municipal library.

THREE GREAT GÖTTINGEN PROFESSORS

A LIST of the mathematicians, starting with the eminent Carl Friedrich Gauss, who either studied or taught at Göttingen University, reads like an honor roll of the profession. The next eighty-four items will be stories and anecdotes about members of this list. When we add to these stories the couple of dozen just told about Gauss, we obtain over a hundred items devoted (in the present work) to mathematicians on the Göttingen list. We here commence with three giants of the list, Georg Bernhard Riemann (1826–1866), Hermann Minkowski (1864–1909), and Edmund Landau (1877–1938).

212° *An interesting analogy.* It is sometimes challenging to search out two men, A and B, in different fields of work about which one can say, with some degree of sense, that A in his field was like B in his. These analogies can be interesting, though they are usually quite limited. An analogy of this sort has been stated between Georg Bernhard Riemann, the mathematician, and Samuel Taylor Coleridge, the poet. Each published surprisingly little in his field, but what each did publish was of a quality worthy of being bound in gold, and each had a considerable influence on the work of others. Of course there are also many contrasts between these two men.

213° *Presents.* One of Riemann's greatest joys as a boy was to make or buy small remembrance gifts for his parents, brother, and sisters. He once presented his parents with an original perpetual calendar that he had invented and that astonished his schoolmates.

214° *Rapid assimilation.* Riemann devoured advanced mathematics texts with ease and with speed. In six days he read and completely mastered the 859 pages of close reasoning found in Legendre's *Théorie des Nombres*.

215° *Contrast.* Many a person who is very shy in his public behavior is very bold in his creative thinking. Riemann was such a person—in his physical life very bashful, timid, and diffident, but in his scientific life very daring, intrepid, and powerful.

216° *Riemann's Habilitationsschrift.* One of the most interesting stories about Riemann concerns his *Habilitationsschrift*, or probationary lecture—a trial lecture that he had to deliver and have accepted before he could be appointed to an unpaid lectureship at Göttingen University. According to custom, he had to submit titles for three different lectures, from which the faculty was to choose one. Expecting that the faculty would follow general practice and pick the first, or possibly the second, of the submitted titles, Riemann assiduously prepared himself for these possibilities, and spent almost no time on the third and rather incautiously submitted title.

Now the third lecture concerned itself with the hypotheses which lie at the foundations of geometry, a topic on which Gauss had pondered for sixty or more years. Gauss was so curious to hear what the brilliant young Riemann might say on this topic that he designated the third lecture as the one for Riemann to deliver. And so, after a frantic eleventh-hour preparation, Riemann presented to the Göttingen faculty, and thus to the world, his third lecture, a masterpiece in both mathematics and exposition, and since ranked as perhaps the richest single paper of comparable length in the history of mathematics. Omitting sticky technical details, so as not to discourage the non-specialists of the faculty who were present, this paper of Riemann's cast out an enormous number of new and highly fruitful ideas, ideas that engaged the attention of researchers for many years to follow.

In the concluding comments of his lecture, Riemann apologized for presenting such an apparently useless topic, but, he said, the value of such an investigation perhaps lies in its ability to liberate us from preconceived ideas should the time ever come when explorations of physical laws might demand some geometry other than the Euclidean. These highly prophetic words were actually realized some fifty years after his death, through Einstein's general theory of relativity.

217° *Genius and craftsmen.* In his splendid book, *Space Through the Ages*, Cornelius Lanczos, speaking of the remarkable geometrical achievements of Gauss and Riemann, says, "The genius illuminates with a single vision a landscape as by sheet lightning. Then the craftsmen rush in and develop the territory to a habitable place."*

218° *Intellectual tonnage.*   Riemann possessed prodigious mathematical gifts, and though he wrote comparatively little, that little was so rich in new ideas and fertile concepts that it led to an enormous amount of significant research on the part of others. Cornelius Lanczos put it well when he said, "Although Riemann's collected papers fill only one single volume of 538 pages, this volume weighs tons if measured intellectually. Every one of his many discoveries was destined to change the course of mathematical science."*

219° *A prodigy.*   Hermann Minkowski was one of the most likeable of all mathematicians. He was uniformly kind in his relations with his students and with his colleagues, and apparently never entertained a mean thought about anyone. He was also powerfully creative—and he was a prodigy. His sister Fanny has told us that even as a schoolboy his mathematical abilities were very impressive, and that whenever his teacher became stuck at the blackboard with a mathematical problem, the fellow students would chorus, "Minkowski, help!"

When Minkowski was a young university student in Berlin, he won a monetary prize for his mathematical ability, but he quietly gave it up to a more needy classmate. Even his family was unaware of this incident, and only learned of it later when the classmate's brother told of it.

When Minkowski was eighteen years old, he won, jointly with the fifty-seven year old Irish mathematician Henry John Stephen Smith, the much-coveted Grand Prix des Sciences Mathématiques of the Paris Academy for his work connected with the representation of a number as the sum of squares. He won even though, because of lack of time, he had not first translated his paper into French, which was

---

* Quoted by permission of Academic Press, New York.

one of the rules of the competition. Smith unfortunately died two months before the Academy made the award.

220° *At a department meeting.*   At a department meeting, Klein once completely filled a large blackboard with data and statistics concerning the German middle schools—among which he was trying to institute an educational reform. Turning to his colleagues he asked if any of them had a question. "Yes," replied Minkowski softly, "Doesn't it seem to you that there is an unusually high percentage of primes among those figures?"

221° *A rare burst of arrogance.*   Minkowski was a gentle and modest man who seldom permitted himself to be in any way arrogant. On one occasion, however, he chanced in a course on topology to mention the four-color conjecture. He told the class that the conjecture was unproved, but only because mathematicians of inferior rank had worked on it; he felt he himself could prove it. And right there in class he started an attempt. He continued his efforts the next time the class met, and for several weeks he struggled in class with the recalcitrant conjecture. Finally, on a rainy morning, just as he entered the classroom there occurred a great clap of thunder. He faced the class and in deep seriousness said, "The heavens are angered at my arrogance. I, too, am unable to prove the four-color conjecture." The four-color conjecture is today still unproved.

222° *Minkowski on Einstein.*   Minkowski used to tell his students at Göttingen that Einstein's presentation of the theory of relativity was mathematically awkward. "I can say this," he said, "because Einstein received his mathematical education at Zürich from me."

223° *Minkowski on Mrs. Hilbert.*   Many of David Hilbert's papers were written out for the publisher in the neat and very readable handwriting of Mrs. Hilbert, and also many of these papers were then checked and proofread by Hilbert's gracious friend Hermann Minkowski. In one of his congratulatory letters to Hilbert, Minkowski wrote, "I also congratulate your wife on the fine example that she has set for all mathematicians' wives, which will now for all time remain preserved in memory." He also accused Hilbert of omitting an important person from his list of acknowledgments. "That you omitted thanks to Mrs. Hilbert I find scandalous, and feel that this simply must not be allowed to remain so."

224° *Minkowski's death.*　On a Sunday afternoon, after dinner, Minkowski was suddenly stricken with a ruptured appendix, and that night it was realized that an operation had to be performed. The next day his condition grew worse. He knew the hopelessness of his situation and tried, on his hospital bed, to correct the proof-sheets of his latest work. He expressed to Hilbert his regrets at his fate, for he had hoped to accomplish so much more. He was also concerned that now he would miss Hilbert's lecture on the Waring problem at the coming meeting of the mathematics seminar. Tuesday noon, January 12, 1909, Minkowski asked for his family and Hilbert. When Hilbert arrived at the hospital the gentle Minkowski was dead.

225° *The spoiled child of mathematics.*　Professor Edmund Landau of Göttingen University came from a rich banking family and had been raised with all the luxuries that his parents could provide. He was a small, chubby man, with an innocent face, a small bristling mustache, and an absolutely complacent self-esteem. When asked how to find his home in Göttingen he would artlessly say, "You will have no difficulty finding it; it is the finest house in town."

226° *A Landau game.*　Edmund Landau loved to invite fellow mathematicians to his beautiful Göttingen home. After dinner, in his study, he encouraged them to play games, all highly sophisticated. One of them I remember well. If "meeting a person" means "shaking hands with him," build the shortest chain of mathematicians (or scientists in general) linking, say, Hilbert with Euler, by persons who met each other.

For example: "Hilbert met Gordon, Gordon met Clebsch, Clebsch met Riemann, Riemann met Gauss, Gauss met Kästner, Kästner met Euler," somebody might say. Then others would try to pick holes in it, perhaps by asking, "Did Kästner really meet Euler?"—or by suggesting shorter chains.

Try the game. How many links do you need in a chain of "meeters" to go back to George Washington, Euler, Martin Luther, or Bowditch? "Meeters" may be anybody, not necessarily scientists.

The game can also be played "horizontally," between contemporaries. How many links connect you and, say, Mao Tse-tung, or Albert Schweitzer?—DIRK J. STRUIK

227° *The two sides of the ledger.*　Edmund Landau possessed a certain overbearing arrogance that at times annoyed his colleagues and students at Göttingen, and many feared his sharp wit and ruthless honesty. But all admired his indefatigable diligence and his impersonal devotion to mathematics. Professor Hardy once remarked that though almost everyone is at bottom a little jealous of progress made by others, Landau was surprisingly free of this unworthy emotion. And as a researcher in analytical number theory, Landau has had few equals.

228° *Felix and Serge.*　A student once consulted Landau concerning the quality of a certain piece of amber. The German word for amber is *Bernstein*. Landau didn't think the piece of amber was particularly superior, so he replied, "Felix." You see, there were two mathematicians named Bernstein, Felix Bernstein and Serge Bernstein, and Landau regarded Serge Bernstein as quite the superior of Felix Bernstein.

229° *Fermat's last theorem.*　Landau kept a form for replying to those who sent "proofs" of Fermat's last theorem. The form read: "On Page ——, lines—— to ——, you will find there is a mistake." The task of locating the mistake fell to the Privatdozent.

230° *Irrepressible.*　In the long partnership of collaboration between Hardy and Littlewood, the latter was much the more self-effacing. Once, when Littlewood visited Edmund Landau at Göttingen, the irrepressible Landau said, "So you do exist! I thought you were merely a name used by Hardy for those papers which he didn't think were quite good enough to be published under his own name." (See Item 301° in *Mathematical Circles Revisited.*)

231° *Slightly incorrect.*   Here is a witticism that Jesse Douglas was fond of telling. At Göttingen he once attended a lecture by Landau on Fourier series, and when Landau explained the so-called Gibbs phenomenon, Landau said, "Dieses Phenomen ist von dem englischen Mathematiker Gibbs [he pronounced it Dzjibs] in Yale [he pronounced it jail] entdeckt worden." Only respect for Landau, said Jesse, prevented him from saying, "Herr Professor, what you say is absolutely correct, but for a few things. Erstens, he was not English, but American. Zweitens, he was not a mathematician, but a physicist. Drittens, his name was Gibbs, not Dzjibs. Viertens, he was not in jail, but at Yale. Und fünftens, he was not the one who discovered it."—DIRK J. STRUIK

232° *Landau leaves Göttingen.*   With the rise of German nationalism, many of the Jewish students and faculty began to leave Göttingen. Then, when Adolf Hitler became chancellor of Germany, the universities were ordered to remove all full-blooded Jews from teaching positions, and the great exodus of Jewish talent really got under way. Landau, because of his great wealth and financial holdings, couldn't bring himself to leave the country. He tried to continue lecturing, but when he announced a calculus course an unruly mob kept him from the lecture hall. So he, too, finally left Göttingen. Professor Hardy arranged for him to give a series of lectures in England. "It was quite pathetic to see his delight when he found himself again in front of a blackboard and his sorrow when his opportunity came to an end," reported Professor Hardy.

233° *Mrs. Landau about her husband.*   Several years ago, at the University of Pennsylvania, Professor Isaac Schoenberg invited a number of graduate students to his home for a buffet supper. While we were there, Mrs. Landau, who was visiting him at the time (Landau's daughter, then deceased, was Schoenberg's first wife) came down and was introduced to us. She had heard some of our conversation regarding problems we were struggling with, and innocently remarked, "You know, my husband was something of a mathematician, too," which has seemed to me to be the understatement of all time.

JOHN C. MAIRHUBER

THE MASTER

DAVID Hilbert (1862–1943), the greatest mathematician of recent times, achieved a perfect equilibrium between intuition and logic, between individual concrete problems and general abstract concepts, between pure mathematics and applied mathematics, and between mathematics and science.

Probably each of us who works at mathematics has a special idol in our field whom we admire above all others. David Hilbert has long been the special idol of the writer of these pages. It therefore particularly pleases this writer to sing a few praises of the recent and truly excellent biography of his hero, *Hilbert*, by Constance Reid (Springer Verlag, 1970)—a brilliant model for subsequent writers of lives of mathematicians to emulate. Our little book of stories will be a success if it does little else than bring this beautifully touching biography to the attention of some readers. Many of the stories we here tell about Hilbert are also told, sometimes in a slightly different version, in Constance Reid's wonderful book.

Highly gifted and highly versatile, David Hilbert radiated over mathematics a catching optimism and a stimulating vitality that can only be called "the spirit of Hilbert"—may that spirit never pass away! Engraved on a stone marker set over Hilbert's grave in Göttingen are the Master's own optimistic words:

> Wir müssen wissen.
> Wir werden wissen.

234° *Stimulation.* Walking seems to be highly conducive to the thought processes, and many mathematicians have arrived at their greatest ideas while walking. Not only has it been common for pairs, or small groups, of mathematicians to ramble about the countryside together, discussing mathematics, but many mathematicians compose mathematics in their study rooms while pacing back and forth.

Hilbert claimed he worked best out-of-doors. He accordingly attached an eighteen-foot blackboard to his neighbor's wall and built a covered walkway there so that he could work outside in any weather. He would intermittently interrupt his pacing and his blackboard computations with a few swirls around the rest of the yard on his bicycle, or he would pull some weeds, or do some garden trimming. Once, when a visitor called, the maid sent him to the backyard and advised that if the master wasn't readily visible at the blackboard to look for him up in one of the trees.

235° *Hilbert conducts business with the Minister of Culture.*
Erich Hecke (1887–1947), who became an accomplished mathematician, served in his early Göttingen days as Hilbert's assistant. Though he was awarded only the paltry sum of 50 marks a month (about $12.50 in United States money at the time) for this work, he always regarded his assistantship with Hilbert as the high point of his life.

Hilbert felt that Hecke should receive greater recompense, and he told Hecke that the next time he went to Berlin he would take up the matter with the Minister of Culture. The occasion soon arose, and after Hilbert had finished discussing university business with the Minister, he realized there was something he had forgotten. He accordingly poked his head out of the Ministry window and shouted down to his wife waiting in the park below, "Käthe, Käthe! About what else did I want to talk to the Minister?" "Hecke," Mrs. Hilbert called back. Hilbert pulled in his head and turning to the startled Minister demanded that Hecke's salary be doubled. It was.

236° *The Wolfskehl Prize.* When the Darmstadt mathematics professor Paul Wolfskehl (1856–1906) died, he bequeathed 100,000 marks to the Göttingen Scientific Society as a prize for the first complete proof of Fermat's last "theorem." Until the prize should be awarded, the interest accruing on the sum was to be used as recommended by a committee of the Society. Hilbert, who became chairman of the committee, was able to arrange that 2500 marks of the interest be used to bring Henri Poincaré to Göttingen for a series of lectures. Hilbert later arranged to bring other guest lecturers in the mathematical sciences to Göttingen in this same way. Someone once asked Hilbert why he didn't try to win the Wolfskehl Prize by proving Fermat's last "theorem." "Why kill the goose that lays the golden eggs?" he replied.

The spiraling inflation in Germany following the disasters of World War I finally rendered the Wolfskehl Prize worthless.

237° *Inflation.* After World War I, the German mark declined ever more rapidly in value as inflation soared exponentially. The situation is strikingly illustrated in the growing cost of a volume of the *Mathematiche Annalen.* Constance Reid, in her book *Hilbert*, reports that in 1920 the price was 64 marks; at the beginning of 1922 it was 128 marks; toward the end of 1922 it was 400 marks. By 1923 it reached 800 marks, and by the end of 1923 it was 28,000 marks.

Inflation ended abruptly in 1923 when the German government created a new unit of currency called the Rentenmark. Hilbert skeptically commented that one cannot solve a problem by merely changing the name of the independent variable.

238° *Requirements for a good problem.* At the International Congress of Mathematicians held in Paris in the summer of 1900, Hilbert listed and discussed twenty-three important and, at that time, unsolved problems in mathematics, the solution of any one of which he felt would materially add to the development of mathematics in the coming century. He prefaced his discussion of the problems with some brief remarks as to what constitutes a good mathematical problem. Such a problem, he said, had to be: (1) clear and easily comprehended, (2) difficult yet not hopelessly inaccessible, (3) fruitful.

239° *Catching a fly on the moon.* Hilbert was asked what technological achievement he would regard as the most important "To catch a fly on the moon," he replied. When the surprised questioner asked why, Hilbert pointed out that the attendant technical problems that would have to be solved first, before a fly could be caught on the moon, would involve solutions of almost all the material difficulties of mankind.

240° *In defense of Galileo.* Someone once blamed Galileo for recanting before the Inquisition (see Item 156° of *In Mathematical Circles*). Hilbert objected: "Galileo was no idiot. Only an idiot could believe that science requires martyrdom—that may be necessary in religion, but in time a scientific result will establish itself."

241° *Hilbert's "flames."* Hilbert had a fondness for pretty young ladies and he flirted outrageously with them in boyish innocence. He was adept at gallant bon mots and grew flowers in his garden for his "flames." He particularly enjoyed dancing with the pretty ladies and it was common at a dance to see him whirling the beautiful young wife of a colleague about the dance floor, unabashedly sweeping her into his arms at the end of the dance and planting a kiss on her lips.

Once during one of his birthday parties he sat in an adjoining room with his arms about the two young nurses who daily gave him physical treatments. When he was urged to come out and listen to the glowing speeches being made about him, he said he was enjoying himself much better where he was.

At another birthday party (his fiftieth), the students decided to honor him with a Love Alphabet—for each letter of the alphabet they were to compose a rhyming couplet about one of the professor's "flames" whose name started with that letter. When they got to the letter K, no one could think of a name. Finally Mrs. Hilbert, whose name was Käthe, pointed out that surely they could just once think of her. The delighted students thereupon composed the couplet

> Gott sei Dank, nicht so genau
> Nimmt es Käthe, seine Frau.

David Hilbert had a remarkably understanding wife.

242° *Brutally direct.* Hilbert could be brutally direct. If one of his students presented something too easy he would cut in with, "But that is completely elementary." If another student gave an inadequate presentation, he would chastize, "Your report has been nothing but chalk, chalk, chalk!" Again, if a student should be long-winded, he would say, "Only the raisins of the cake, please."

Of his physicist colleagues he said, "Physics is much too difficult for physicists." He once asked someone if a certain mathematician was still alive. The person said yes, and then went on to tell where the man was living, what he was teaching, what he was researching on, about the man's family, and so on. Hilbert finally managed to interrupt with, "Yes, but I don't want to know all that; I only asked, *does he still exist?*" Once, upon being asked what he thought of astrology, he replied, "If you should assemble the ten wisest men of the world and ask them to find the most stupid thing in existence, they will be unable to find anything more stupid than astrology."

243° *Hilbert's harshness.* Hilbert could be rudely harsh. Thus there was the occasion when he interrupted a speaker at the Mathematics Club with, "My dear colleague, it is apparent that you don't know what a differential equation is." The speaker, shocked and humiliated, stalked out of the room and into the adjoining mathematics library. Thereupon, many who were present reproached Hilbert for the rawness of his remark. "But it is clear that he doesn't know what a differential equation is," Hilbert countered. "See, he has now gone into the library to look it up."

On another occasion Norbert Wiener, then a young man, gave an important talk at the Mathematics Club. After the meeting, the group hiked up to Der Rohns for their customary supper together. During the supper Hilbert rambled on about talks he had heard at Göttingen. The more recent ones, he said, were much poorer than those of some years ago, and those of the past year had all been especially bad. "Except for the one this afternoon," he said. Then, as Wiener prepared himself for a compliment, Hilbert went on, "The talk this afternoon was the very worst we have ever had."

244° *Geheimrat Hilbert.* Someone, who kept addressing Hilbert over and over as "Herr Geheimrat," noticed that the professor looked irritated, and worriedly asked, "Am I annoying you, Herr Geheimrat?" "There is nothing about you that annoys me," replied Hilbert, "except your obsequiousness."

245° *Poor memory and slow understanding.* Throughout his life, Hilbert was poor at memorization and was slow at grasping new ideas presented by others. He later claimed that mathematics first appealed to him because it required no memorization—he could always figure things out for himself. And as for any new idea, to understand it properly he had to work it all out for himself from the beginning. At mathematical gatherings he astonished many quick minds by his slowness at comprehending new ideas.

246° *Hilbert's obtuseness.*   Hilbert was quite slow at grasping new or complicated concepts in mathematics, and the young men attending the meetings of the Göttingen Mathematics Club were frequently surprised at Hilbert's difficulty over things that they had no trouble in seeing immediately. Hilbert, looking perplexed, would interrupt the speaker for clarification, and it was not uncommon finally for all present to join in trying to help him understand. Hilbert himself confessed that whenever he read or was told something in mathematics, he found it very difficult, indeed often almost impossible, to understand the explanation. He had to seize the thing, get to the bottom of it, chew on it, and work it out on his own—usually in a new and much simpler way.

247° *Hilbert's class lectures.*   Though Hilbert was a very inspiring lecturer, he frequently garbled things at the blackboard. He only prepared an outline for his lectures and left the details to be worked out during delivery in the classroom. It was not uncommon for him to get stuck at the blackboard even in his elementary courses. In such cases he would drop the matter with a flourish of his hand and a comment that it was trivially easy. Fortunately he was blessed with outstanding assistants who, after taking notes of the lecture, would then proceed to smooth out the muddled parts so that the students could later pick up good clean copies of the material.

248° *Existence.*   To illustrate a claim of the existence of something by merely asserting a purely existential statement rather than by actually producing the specified object, Hilbert used to say to his class, "Among the people now in this lecture hall, there is one who has the least number of hairs on his head." This always brought a laugh, presumably because the individual he referred to was quite clearly himself.

249° *Hilbert and the new quantum mechanics.*   I attended Hilbert's seminar. One afternoon he came running in and bawled us out: "There you are sitting and talking about your petty problems! I have just come from the physics seminar—und da machen sie gerade die groszartigsten Sachen!"

The year was 1925; the seminar was Max Born's. Heisenberg had just been lecturing on his new quantum mechanics.

DIRK J. STRUIK

250° *Hilbert and the prime numbers.* When, in 1925, I attended Hilbert's class at Göttingen on number theory, he began by writing down all the prime numbers below 100. The next hour, before he continued his lecture, he came running into the room. "Ach, I have forgotten one prime number! Das darf nicht sein, sie sind schön—man muss sie gut behandeln!"—DIRK J. STRUIK.

251° *Student adulation.* There were occasions when other institutions endeavored to entice Hilbert away from Göttingen. This happened when Lazarus Fuchs died in 1902 and his chair at the University of Berlin was offered to Hilbert.

When the news of the "call" to Berlin became known to the mathematics students at Göttingen, they became upset and worried that Hilbert might accept. Finally a delegation of three students, headed by Walther Lietzmann (who himself later became a distinguished mathematician), was selected to go to Hilbert's home and express the deep wish that he remain at Göttingen. Hilbert listened to the students but made no comment, and after a drink of punch served by Mrs. Hilbert, the delegation left feeling quite discouraged.

Hilbert seemed slow coming to a decision, and his frequent trips to Berlin and the obvious fact that his mind was not entirely on his teaching, led everyone to believe he intended to accept the Berlin post. Actually, however, Hilbert was planning how to use the Berlin offer to improve the department at Göttingen. Finally, as his price to stay at Göttingen, he secured permission to create a new professorship in mathematics at Göttingen and to invite Minkowski to fill it. The whole maneuver was carried out with great diplomatic skill.

When the students and the members of the Göttingen Mathematics Club finally heard the good news that Hilbert was staying and also that Minkowski was coming, they celebrated by organizing a great smoking and drinking party in Hilbert's honor, at which Felix Klein delivered a magnificent speech praising Hilbert's work in mathematics.

252° *Hilbert and Ackermann.* Hilbert was against young mathematicians and scientists marrying; it prevented them from fulfilling their obligations to their subject. When Wilhelm Ackermann (1896–1962), who had collaborated with Hilbert on an excellent book on logic, married, Hilbert became very angry and he refused to assist Ackermann further in his career. Unable to obtain a university post without Hilbert's help, the gifted young man had to take a position teaching in a high school. When Hilbert later learned that the Ackermanns were expecting a baby, he exclaimed, "That is wonderful news, as it excuses me completely from having to do anything more for such a crazy man."

253° *The Riemann hypothesis.* There is a story, probably apocryphal, to the effect that one of Hilbert's students once submitted a purported proof of the Riemann hypothesis. Hilbert was much impressed with the effort, even though he found it contained an error. Not long after, the student died, and Hilbert secured permission from the parents to speak a few words at the boy's funeral. As the parents and some friends stood at the graveside in the rain, Hilbert stepped forward. He started by lamenting the tragedy of the gifted boy's early death. He went on and remarked that though the young man's proof of the Riemann hypothesis contained an error, it was quite possible that someone might later secure a sound proof along the lines laid down by the deceased. "In fact," he said, with a sudden light in his eyes, while the rain continued to fall at the graveside, "let $f(z)$ be a function of the complex variable z. Consider. . . ."

254° *Hilbert and Hilbert space.* Many a mathematical concept has been generalized and extended so far beyond its original substance that the motivating idea has become scarcely recognizable. This happened to a brilliant idea put forth by David Hilbert, wherein he ingeniously placed geometry in the service of analysis by a remarkable isomorphism which he established between the geometrical theory of quadratic surfaces in Euclidean space of many dimensions and properties of linear differential and integral operators. These geometrical ideas of Hilbert were later developed into a considerably more abstract and sophisticated formulation that received the name *Hilbert space,* so christened in 1931 by John von Neumann in an important paper of his devoted to the mathematical foundations of quantum mechanics. Schrödinger's earlier work in quantum mechanics had given particular weight to Hilbert's original idea, but the subsequent demands of advanced quantum theory required von Neumann to formulate a more abstract approach which broke through some of the limitations of Hilbert's geometrical picture. Thus Hilbert's basically geometric discussion was retranslated into a much more analytical and formalistic language while still retaining a good deal of the former geometrical terminology. A story is told that after his retirement Hilbert once attended a seminar at Göttingen in which some mathematical consequences of Hilbert space were discussed. At the conclusion of the lecture, a perplexed Hilbert rose and asked, "Sir, can you please tell me, what is Hilbert space?"

255° *Taking credit.* As a boy, David Hilbert's son Franz did not take much to mathematics. Hilbert used to say, "He inherited his mathematical ability from his mother; everything else he got from me."

256° *Franz Hilbert.* There is something pathetic about Franz Hilbert, the only child the Hilberts had. He was born in the summer of 1893 at the seaside vacation resort of Cranz. In his early youth his father entertained some hopes for him, but, though not unintelligent or untalented, the boy seemed highly strung and at times confused. In school he appeared to have very little retentive power and accordingly did poorly. Later, when the army did not take him, he tried a number of small jobs, at all of which he failed. He seemed increasingly disturbed. Then one day he had a tragic breakdown and had to be admitted into a mental institution. At this, Hilbert disowned his son and thereby increased Mrs. Hilbert's sorrow. Franz was later released from the hospital and tried further little jobs, but he could not keep any of them very long. As he aged he came to look more and more like his famous father, many of whose mannerisms he endeavored to imitate, but without the accompanying mental force of his father the imitation was only a tragic parady. He frequently spoke of learning mathematics so that he might be able to appreciate his father's work. When Mrs. Hilbert died in 1945, almost blind and a couple of years after the death of her husband, there was no old friend available to speak beside her coffin. Franz finally begged a woman who had never known the family well to say a few words. Franz himself lived a long life, passing away in 1969 at the age of seventy-six.

There is disappointment when one learns that one's mathematical hero was at times much less of a hero in his personal life.

257° *Imagination.*   Once, when a mathematician had turned novelist and exclamations of surprise arose as to how such a thing could happen, Hilbert said it was easy—the man lacked sufficient imagination to be a mathematician but had enough to be a novelist.

258° *Point of view.*   On another occasion Hilbert said that sometimes a person's circle of broadmindedness becomes smaller and smaller, and as the radius approaches zero and the circle approaches a point, that single point becomes the person's point of view.

259° *The frog and mouse battle.*   There were occasions in the battle between the intuitionists, led by L. E. J. Brouwer, and the formalists, led by David Hilbert, when the participants became over-heated.

When Brouwer was invited to the Göttingen Mathematics Club to give a talk on his ideas, someone objected, "You say that we cannot claim that the statement, 'In the decimal expansion of π the sequence of digits 0123456789 occurs,' is true nor that it is false. Perhaps we cannot, but God can." "I have no pipeline to God," replied Brouwer acidly. Finally Hilbert arose. "According to your tenets," said Hilbert to Brouwer, "most of the results of modern mathematics must be abandoned; to me the important thing is not to get fewer results in mathematics, but to get more results."

Brouwer became a fanatic for his cause and considered Hilbert as his personal enemy. He once stalked out of a gathering when van der Waerden, who was also present, referred to Hilbert as his friend.

Both Brouwer and Hilbert were on the editorial staff of the *Mathematische Annalen.* Hilbert was one of three chief editors and Brouwer was a member of a seven-man board of associate editors. Brouwer demanded that all papers submitted to the *Annalen* by Dutch mathematicians, and all papers in topology, be given to him to referee. The other editors objected to this dictatorial demand. Hilbert, fearing for the integrity of the *Annalen*, felt that Brouwer must be removed from the editorial staff. He further feared that if, for reasons of health, he himself had to withdraw from the journal, Brouwer might take over— to the detriment of the journal. Since it was difficult to ask Brouwer alone to resign, Carathéodory, who was on the seven-man board, recommended that the entire seven-man board be dismissed. Hilbert promptly adopted the suggestion.

Albert Einstein, who had been one of the three principal editors, became so disturbed by the controversy that he resigned. "What is this frog and mouse battle among the mathematicians?" he wondered.

260° *Hilbert on the law of excluded middle.* One of the tenets of the intuitionist school of philosophy in mathematics is that the law of excluded middle does not hold universally—it holds in finite situations but not in infinite ones. Now in mathematics, the law of excluded middle is often powerfully employed in the prohibited fashion. Hilbert declared that "to prohibit a mathematician from using the law of excluded middle was like prohibiting a boxer from using his fists."

261° *No roast goose.* During World War I, when food, especially meat, had begun to be scarce in Germany, Hilbert, with other Göttingen academic barons, was called to an evening gathering in the Great Hall by the University Rector. The last time a similar gathering had been called, the university had distributed some succulent geese among the chosen of the faculty. Hilbert now convinced himself that the new occasion was for a similar reason, and he primed himself for the anticipated gift of more geese, or perhaps some other choice meat. When the group assembled, it was found that there was no meat to be handed out; the Rector merely wanted to announce with great ceremoney that Kaiser Wilhelm had just declared unrestricted submarine warfare on the enemy. Amidst the cheers of the other professors, Hilbert, disappointed and disgusted, turned to a neighbor and said, "The German people are like that. They don't want geese; they prefer unlimited submarine warfare."—B. L. van der Waerden.

262° *Memorial to Darboux.* In 1917, the same year in which the United States entered the war against Germany, the great French mathematician Gaston Darboux (1842–1917) died. Hilbert, who had admired Darboux, immediately set about preparing a memorial tribute for publication in the *Nachrichten*. When the tribute appeared, a mob of incensed students marched to Hilbert's home and demanded that the memorial to the enemy mathematician be disavowed and withdrawn. Hilbert refused and then threatened to resign from the university if he did not receive an official apology for the students' barbaric behaviour. He received the apology and his memorial to Darboux remained in print.

263° *Declaration to the Cultural World.* In an effort during World War I to convince the enemy that cultural Germany supported military Germany, a Declaration to the Cultural World was drawn up· and passed around to certain great scientists, artists, and writers to sign Here it was declared that Germany had not caused the war, and the "slanders and lies of the enemy" were listed and denied. Felix Klein, who was an ardent patriot, signed the declaration without carefully reading it, somewhat to his dismay later when he took time to examine what he had put his name to. Hilbert, who also was asked to sign, refused, and many thereby regarded him as a traitor. Another conspicuously absent signature was that of Albert Einstein, then at the Kaiser Wilhelm Institute in Berlin. But Einstein had become a Swiss citizen, and so escaped the opprobrium of traitor. Klein was dropped from membership in the Paris Academy; Hilbert was permitted to remain.

The whole war, with its growing anti-Semitism, was completely baffling to Hilbert. He was never able really to understand it. He could not account for such seemingly aberrant behavior on the part of supposedly civilized men.

264° *A medical success story.* In his early sixties, Hilbert showed steadily growing signs of ailing, and by the fall of 1925 his trouble was finally diagnosed as pernicious anemia. There was no known treatment for this disease at the time, and Hilbert's doctors predicted he could live only a few more months. But a pharmacologist friend at Göttingen happened to read in the *Journal of the American Medical Association* about work being done in connection with pernicious anemia by G. R. Minot, wherein the patient's blood is regenerated by feeding the patient quantities of an extract of raw liver. Mrs. Landau, who had contacts with the medical profession, with the aid of Richard Courant, sent a long telegram to Minot, who was then at Harvard, requesting aid for the ailing Hilbert. Another telegram was sent to Professor Kellogg, of the mathematics faculty at Harvard, to follow up the request.

Minot and his associates were not overly receptive—there was very little of the liver extract on hand, it had to be taken by the patient for the rest of his life, and there were people right there in America dying from pernicious anemia. Professor George Birkhoff, also of the Harvard mathematics staff, had recently seen Shaw's play, *The Doctor's Dilemma,* in which a doctor able to save only ten lives must decide which lives he should save. Birkhoff quoted from the play to Minot, and Minot gave in.

A telegram was sent to the Göttingen pharmacologist, instructing him how to prepare raw liver to serve for treatment until the extract could come later. Hilbert declared he would rather die than eat raw liver, but eventually the extract arrived. Hilbert's condition improved immediately, and he outlived his doctors' original grant of only a few months by close to twenty years.

265° *An anti-Semite.* The mathematician Ludwig Bieberbach was very anti-Semitic. Along with others, he tried to analyze the differences between German and Jewish mathematics, in an effort to make the latter appear inferior.

There was a joke at Göttingen that the only Aryan mathematician at the university there really had Jewish blood in his veins; when Hilbert was suffering from pernicious anemia he had received a blood transfusion from Richard Courant.

266° *Bieberbach and the Second International Congress.* Bieberbach was a passionate German nationalist and strongly opposed Germany sending a delegation to the International Congress of Mathematicians held at Bologna in 1928 after the war was over. Backed by Brouwer, who, though Dutch, supported the German nationalists, Bieberbach sent letters to all the secondary schools and colleges in

Germany urging a boycott of the congress. Hilbert countered with a letter of his own, pointing out the folly of Bieberbach's proposal. He said it could only result in misfortune for German science, and expose Germany to justifiable criticism on the part of others.

267° *Hilbert and the Second International Congress.* Though somewhat shaky in health, in August of 1928 Hilbert led a delegation of sixty-seven German mathematicians to the International Congress of Mathematicians held at Bologna, Italy. It was the first time since the War that German scientists attended an international meeting. The delegation was given a standing ovation by all present, and Hilbert commented, "I am very pleased that after a long and arduous time, mathematicians of all countries are gathered together. So it should be, and so it must be, for the progress of our beloved science. . . . To form differences according to people and races is a complete misunderstanding of our science, and can be defended only by very shabby reasons. Mathematics knows no races or geographic boundaries; for mathematics, the cultural world is one country."

There is pleasure when one learns that one's mathematical hero was also at times as fine a hero in his personal life.

268° *Hilbert's hotel bill.* When, on leaving the Second International Congress, Hilbert proceeded to settle his hotel bill, he was told that his accommodations had already been taken care of by the committee that planned the Congress. "Ah, if only I had known that," he said, "I would have eaten a lot more."

269° *Hilbert Strasse.* When Hilbert retired from teaching in 1930, a street in Göttingen was named Hilbert Strasse after him. "Isn't that a thoughtful idea, David," asked Mrs. Hilbert of her husband, "to name a street after you?" "Not the idea, but the act," replied Hilbert. "Klein died before they named a street after him."

270° *A comparison of two deaths.* When Hilbert died, Max Born, who had been at Göttingen when Minkowski died, was in war-bombed England, and he wondered, "Hilbert outlived his friend Minkowski by over thirty years, and was permitted to go on accomplishing important work. But who would care to say if his lonely death in the dark Nazi days was not even more tragic than Minkowski's at the height of his power?"

QUADRANT FOUR

From a student rationale
to the wheels of thought

FURTHER GÖTTINGEN MATHEMATICIANS

HERE are stories about some further mathematicians who either studied or taught at Göttingen University. Felix Klein headed the mathematics department there for many years.

271° *Ordinary and extraordinary.* In Germany an *extraordinary* professor is merely an assistant professor, and is usually considered in faculty circles as quite inferior to an *ordinary*, or full, professor. The students at Göttingen had their own explanation: "An extraordinary professor knows nothing ordinary and an ordinary professor knows nothing extraordinary."

272° *Felix the Great.* Felix Klein (1849–1925) on occasion gave dinners in his home to which students and colleagues were invited, and he presided over these dinners with a very kingly air. So awesome was the host that it is said a student would stand up when the presiding potentate directed him a question.

273° *A distant god.* Felix Klein took himself and his many projects very seriously. It was said that he allowed himself two jokes, one for the fall semester and one for the spring semester. He budgeted his time so carefully that his own daughter had to make an appointment to talk with him.

274° *A faux pas.* The title "Geheimrat" is one of the most highly respected of titles in German science, and is somewhat equivalent in England to a scientist who has been knighted. Both Hilbert and Klein had received the title of Geheimrat; Hilbert put little value on it, but Klein insisted upon being addressed by it. When Norbert Wiener once called on Klein he committed a terrible blunder. Met at the door of Klein's house by an elderly housekeeper, he asked, "Ist der Herr Professor zu Hause?" With stern rebuke in her voice, the housekeeper replied, "Der Herr *Geheimrat* ist zu Hause."

275° *A syllogism.* With advancing age, Klein became more and more Olympian, and a favorite syllogism circulated among the students at Göttingen: "There are two kinds of mathematicians at Göttingen— those who do what they want but not what Klein wants, and those who do what Klein wants but not what they want. Klein is clearly of neither kind. Therefore Klein is not a mathematician."

276° *Runge as a calculator.*   In the winter term of 1904–1905, the distinguished mathematician-physicist Carl Runge (1856–1927) joined the Göttingen faculty. Soon the weekly Thursday walking trio of Klein, Minkowski, and Hilbert became a quartet. Runge possessed an impressive gift for computation which amazed even his new colleagues. On one occasion, when they were endeavoring to set a schedule for a conference to be held several years in the future, it became necessary to know the date of Easter. Now the date of Easter, depending as it does on the phases of the moon, is no simple thing to determine, and the mathematicians began to look for a calendar. But Runge stood thoughtful for a moment and then announced the required date to his astonished friends.

277° *Runge introduces skiing.*   Runge, who became the sportsman of the Göttingen faculty, introduced skiing to the college community. Hilbert and some of the younger instructors decided to learn the strenuous but enjoyable sport under Runge's coaching.

One day at the weekly meeting of the Göttingen Mathematics Club, Hilbert told his friend Minkowski of a plight he had gotten into while skiing earlier that day. He said he had fallen into a ditch and found himself on his back with both legs up in the air. Right then one of the skis fell off and slid down the hill. He had therefore to take off the other ski and encumbered with it go down the hill in the deep snow to retrieve the escaped one. And that, Hilbert concluded, was not so simple.

"Why didn't you," asked Minkowski, who was not among the ski enthusiasts, "let the second ski slide down the path of the first one? It would have come to rest alongside of the first one."

"Oh," replied Hilbert, "Runge never thought of that!"

One wonders if Item 303° of *Mathematical Circles Revisited* is perhaps an imperfect recollection of this story.

278° *A proof of an impossibility.* Ernst Zermelo (1871–1956) was a solitary man with a preference for whisky over companions When he was at Göttingen he used to argue that it is impossible for anyone ever to reach the North Pole, because the amount of whisky needed to reach any latitude, he said, is proportional to the tangent of that latitude.

279° *Walzermelodie.* When people asked Zermelo about his unusual name he would explain to them that it used to be *Walzermelodie*, but that later it became necessary to discard the first and last syllables.

280° *No effort too great.* Constantin Carathéodory (1873–1950) was on his way to a promising career as an engineer when, at the age of twenty-six, and quite against his family's advice, he decided to return to school and devote his life to the study of pure mathematics. He felt that only in this way would his life become truly worthwhile. The Carathéodory family was well known and very influential in Greece; the motto of the family was, "No effort too great." Constantin lived up to the family motto, contributed notably to mathematics (particularly in function theory), and produced some classic books in the field.

281° *Blumenthal's dissertation.* Otto Blumenthal (1876–1944), when he was a student at Göttingen, one day sadly discovered that the prettiest part of his dissertation had already been written up in another paper. When he notified Hilbert of this dismaying fact, Hilbert merely shrugged his shoulders and asked, "Why do you read so much literature?"

282° *Blumenthal's end.* In the tragic days of the rise of Hitler, with the help of friends Blumenthal was smuggled out of Germany to Holland. But on one of its periodic raids on the Jews of Holland, the German Gestapo netted Blumenthal and he was sent to Theresienstadt, a Czechoslovakian village that the Nazis had converted into a ghetto-prison for old Jews. At one time Blumenthal was placed on a train for the infamous Auschwitz, but was strangely removed before the train departed. He died in the Theresienstadt ghetto some time toward the end of 1944.

283° *First member of the honors class.* The first of Hilbert's famous twenty-three Paris problems to receive solution was the third on the list. It was solved by one of Hilbert's own students, the 22-year-old Max Dehn (1878–1952), who obtained a partial solution within the year of proposal (1900), and a complete solution in the following year. Thus Max Dehn became the first member of "the honors class" of mathematicians who have either partially or completely solved one of Hilbert's twenty-three problems.

284° *Emmy Noether as a lecturer.* Emmy Noether (1882–1935) was a poor lecturer and lacked pedagogical abilities Her classes, accordingly, were always very small. One day at Göttingen, however, she arrived at her class and was surprised to find that, in addition to the regular students, there were so many visiting students that the total audience exceeded a hundred. After her lecture was over, a note was handed up to her which read, "The visiting students have understood your lecture just as well as have the regular students."

285° *The Noether boys.* Emmy Noether bore both a personal and a professional concern for her students; in Göttingen they were called "the Noether boys "—RORA IACOBACCI

286° *Hilbert and Emmy Noether.* Emmy Noether was already famous, but could not get on the Göttingen faculty because the majority was against admitting a woman. Said Hilbert in the faculty meeting, "Aber meine Herren, die Universität ist doch keine Badeanstalt!" This was in the years when there was no mixed bathing.

DIRK J. STRUIK

287° *Max Born's examination.* Max Born, when he was writing his dissertation as a student at Göttingen, inquired of Hilbert how he might best prepare himself for the mathematics part of his examination. Hilbert asked him in what area he felt most poorly prepared, and Born told him "ideal theory." Upon this, Born assumed that at the examination Hilbert would not ask him any questions on ideal theory, but on the fateful day it turned out that *all* of Hilbert's questions were in this area. When the matter was later brought up, Hilbert said, "I was merely curious to see how much you know about something about which you know nothing."

288° *Twice disappointed.*   The two chief figures at Göttingen in the early days of quantum mechanics were Max Born and his favorite student Werner Heisenberg. Born, a gentle and modest man of Jewish origin, saw Heisenberg attracted to the rising German nationalism and ultimately become a member of the Nazi Party.

After the war, when Born retired to England, his most brilliant student was Klaus Fuchs, who later confessed to the British authorities that as a physicist at Los Alamos working on the atomic bomb project, he had passed on vital information to Russian agents.

289° *The Bohr brothers.*   When the two Bohr brothers, Niels and Harald, were small boys, a friend of the family sympathized with the mother for having two such dull children. Niels (1885–1962) became a national hero of Denmark because of his outstanding work in science, and Harald (1887–1951) became recognized as Denmark's foremost mathematician.

290° *Harald Bohr's examination.*   Harald Bohr enjoys the unique distinction of being probably the only person whose doctoral examination in mathematics (taken at Göttingen) was reported in the sports columns of the newspapers. He was a famous athlete and in the 1908 Olympic games had been a member of Denmark's runner-up soccer team.

291° *Siegel goes to Göttingen.*   Carl Ludwig Siegel refused to serve in the German army during World War I and was accordingly confined in a mental institution. The institution was located next to a clinic owned by Edmund Landau's father. In this way Siegel became acquainted with the eminent number theorist of Göttingen. The result was that in 1919 Siegel became a mathematics student at Göttingen. In time he became one of the world's outstanding number theorists.

292° *A poor prediction.*   Hilbert proposed, as the seventh problem in his list of twenty-three significant unsolved problems given at the

1900 International Congress of Mathematicians, a proof of the transcendence or of the nontranscendence of $2^{\sqrt{2}}$. This problem remained refractory for about thirty years.

In 1920, during a lecture on the theory of numbers, Hilbert wanted to give his audience some examples of problems that may seem simple on a first acquaintance, but which are really very difficult; he chose the Riemann hypothesis, Fermat's last "theorem," and the transcendence of $2^{\sqrt{2}}$. He pointed out that progress had been made on the Riemann hypothesis problem and believed he himself might live to see its solution; some of the young students in his audience, he said, might live long enough to see Fermat's last "theorem" cracked; but no one present, he asserted, would live long enough to see a proof of the transcendence of $2^{\sqrt{2}}$. Carl Ludwig Siegel, as a 23-year-old student at Göttingen, heard Hilbert's lecture, and about ten years later, utilizing some work done by the Russian mathematician A. O. Gelfond (who had managed to establish the transcendence of $2^{\sqrt{-2}}$), succeeded in proving the transcendence of $2^{\sqrt{2}}$.

The Riemann hypothesis problem and Fermat's last "theorem" are today still unproved conjectures!

293° *A paper for the Annalen.*   Carl Siegel was once asked to referee a paper for the *Mathematische Annalen*, and finding the paper both partially incorrect and unnecessarily circuitous he recommended rejection. Hilbert, who was one of the principal editors of the journal, returned the paper to Siegel, saying that he felt obligated to publish the paper since the author was on the committee in 1910 that awarded him the Bolyai Prize. "Fix whatever needs fixing," he asked Siegel. So Siegel corrected the paper and the improved version was published in the *Annalen*. Hilbert said nothing to Siegel, but several months later Siegel received a package containing Minkowski's two-volume collected works, bearing the inscription, "With friendly wishes from the editor."

294° *Van der Waerden as a boy.*   The story is told that when B. L. van der Waerden (b. 1903) was a boy, his father (who was a high school teacher) took away his mathematics books, feeling that the boy should be outdoors playing with other boys. When the father later discovered that his son had invented trigonometry entirely on his own, devising his own names and notations for the trigonometric functions, he restored the books.

295° *How Jacob Grommer was led to mathematics.* There was a Lithuanian mathematics student at Göttingen who took up mathematics for an unusual reason. His name was Jacob Grommer, and he suffered from acromegaly, a chronic nervous disease characterized by an enlargement of the feet and hands. Grommer had prepared himself in a Talmudic school to become a rabbi. But it was traditional in the part of Europe from which he came that the new rabbi marry the daughter of the former rabbi. When the old rabbi's daughter saw Grommer's grotesque feet and hands she refused to marry him. His rabbinical career thus ended, Grommer turned for consolation to mathematics.

Grommer wrote a brilliant paper which, because of a technicality, was considered unacceptable as a doctoral dissertation. The technicality was that since Grommer had never received a gymnasium diploma (he had attended a Talmudic school instead of a gymnasium) he was not eligible for a doctoral degree. Hilbert took up Grommer's case, asserting that if students without a gymnasium diploma should always write dissertations of the caliber of Grommer's, it would become necessary to pass a law forbidding the awarding of the gymnasium diploma. Grommer eventually received his degree of doctor of philosophy.

MORE GERMAN MATHEMATICIANS

FRANZ Ernst Neumann (1798–1895) was the father of the school of applied mathematics at the University of Königsberg, where he originated the format of the academic seminar and established the first institute of theoretical physics in a German university. Hermann Günther Grassmann (1809–1877), of Stettin, played an historical role in the liberation of algebra from its traditional mold. Ernst Eduard Kummer (1810–1893), professor of mathematics at the University of Berlin, became closely identified with number theory and introduced the concept of ideal numbers. Leopold Kronecker (1823–1891), who had Kummer as a teacher in both his gymnasium and his college years, also taught at the University of Berlin and was an accomplished algebraist and the strong advocate for the arithmetization of mathematics. Lazarus Fuchs (1833–1902), another professor at the University of Berlin, specialized in the theory of linear differential equations. Wilhelm Blaschke of Hamburg was born in 1885 and was a powerful geometer and author of several classics in his field. P. Koebe of Leipzig was an analyst.

296° *Reward for a discovery.*   At one time an important academy attempted to arrive at some regulations for the awarding of credit for scientific discovery. The versatile Franz Ernst Neumann, who almost became a centenarian and many of whose discoveries were never made public, said, "The greatest reward lies in making the discovery; recognition can add little or nothing to that."

297° *Discouragement.*   In the year 1844, Hermann Günther Grassmann published the first edition of his remarkable *Ausdehnungslehre* (Calculus of Extension). Unfortunately, Grassmann was a very poor expositor and his obscure presentation remained practically unknown to his contemporaries. A second reformulation, put out in 1862, proved scarcely more successful. The resulting complete indifference on the part of the scientific world so discouraged Grassmann that he gave up mathematics for the study of Sanskrit language and literature, a field in which he contributed a number of brilliant papers.

298° *A Kummer simile.*   Ernst Eduard Kummer, who was a remarkably broad and creative mathematician, had the easygoing nature, lively humor, and bluff simplicity of the conventional German of his day. He became famous in his teaching for his homely figures of speech and his semiphilosophical similes. Once, to emphasize the importance of a particular factor in a certain expression, he remarked, "If you overlook this factor you will be like a man who in eating a plum swallows the pit and spits out the pulp."

299° *Kummer unexpectedly wins a prize.*   Kummer once won a prize for which he had not competed. The French Academy of Sciences had set a competition for the Academy's "Grand Prize" in 1853. Not receiving any competing works deemed worthy of the prize, the competition was continued to 1856. Still receiving only inferior papers, it was finally decided, in 1857, to award the prize to Kummer for his researches on the complex roots of unity—work which Kummer had not submitted to the Academy in competition for the Grand Prize.

300° *Kummer's empathy.*   Anyone who studied under Kummer ever after remembered him as both a great teacher and a warm friend. Many stories are told of his empathy for and interest in his students. Thus there was the time when a needy young student was stricken with smallpox just before he was to take his doctoral examination. The young man was forced to forego the examination and to return to his home in Posen by the Russian border. Hearing nothing from the unfortunate student, and fearing the young man might be unable to afford proper medical assistance, Kummer sought out one of the student's friends, supplied him with money, and sent him off to Posen to see to it that whatever needed doing there was done.

301° *Kummer's last nine years.*   Kummer spent the last nine years of his life in absolute retirement, surrounded only by his family of nine surviving children. He completely abandoned his mathematics and, except for a few trips to the scenes of his boyhood, lived in strict seclusion. He died, after a short attack of influenza, at the good old age of eighty-three.

302° *Lotus-eaters.*   Leopold Kronecker said that mathematicians who work in number theory are like lotus-eaters—having once tasted of this food they can never give it up.

303° *Hot from the forge.*   Lazarus Fuchs of Berlin had one quality in common with J. J. Sylvester: only rarely did either prepare his classroom lectures beforehand, but rather produced the material on the spot. The students of these two great mathematicians had the unusual opportunity of frequently actually seeing mathematics in the making by a master.

304° *Blaschke's bad mark.* Professor Wilhelm Blaschke of Hamburg was an eminent and creative geometer, and he wrote a number of exceptionally fine geometry texts. But he has a bad mark against him; he became an ardent supporter of the Nazis and he wrote articles ridiculing American mathematics. He showed contempt for the great school of mathematics at Princeton and called Princeton "a little Negro village."

305° *Immortality.* The story is told that P. Koebe, of Leipzig and of function-theory fame, on viewing Leonardo da Vinci's badly deteriorated painting of The Last Supper, exclaimed, "How sad! This painting will pass away, while my theorem on the uniformization of analytic functions will endure forever!"

OLLA-PODRIDA

OUR stew will be a mixture of three M's—Mathematicians, Mathematics, and Miscellaneous.

306° *A means of escape.* It has been observed by some that mathematicians seem generally to meet family and personal tragedies more stoically than do many other people, as though through preoccupation with their subject they have somehow acquired a strength to meet the changing courses of life. Minkowski ventured the opinion that this is not the true explanation, but rather that the mathematicians have really done no more than to find in their work an escape from their sorrows.

307° *Depressed mathematicians.* Many great mathematicians have, like Felix Klein and G. H. Hardy, undergone periods of deep depression. Richard Courant sees this as natural, for there must be periods in the life of almost any creative person when he feels he is losing his powers. This comes as a dismaying shock and leads to a fit of depression.

308° *The spirit of mathematical inquiry.* Our knowledge of mathematics and of the real world constantly grows in ever-swelling circles. One is reminded of the engraving of the logarithmic spiral (see Figure 65) along with the motto *Eadem mutata resurgo* ("I shall arise the

FIGURE 65

same though changed") on Jakob Bernoulli's tombstone—see Item 205° of *In Mathematical Circles*. The figure and the motto are symbolic of the ever-widening and yet forever-changing spirit of mathematical and scientific inquiry.

309° *The sure test.* "I constantly meet people who are doubtful, generally without due reason, about their potential capacity [as mathematicians]. The first test is whether you got anything out of geometry. To have disliked or failed to get on with other [mathematical] subjects need mean nothing; much drill and drudgery is unavoidable before they can get started, and bad teaching can make them unintelligible even to a born mathematician."

J. E. LITTLEWOOD

310° *A truism.* "The infinitely competent can be uncreative."

J. E. LITTLEWOOD

311° *Pioneer work is clumsy.* "A mathematician's reputation rests on the number of bad proofs he has given "—A. S. BESICOVITCH

312° *How to win a war.* In these days, a war is won as much by mathematical scientists as by soldiers. Now a good mathematical scientist is, by nature, incurably addicted to puzzle problems. For example, it was estimated that in England during World War II, the intriguing "false coin" problems* wasted something like 10,000 scientist-hours of war work. Why not, then, drop some new, challenging, and difficult puzzle problems over the enemy country, and thus demobilize the enemy's mathematical scientists?

313° *The Greeks versus the European Christians.* Aristarchus of Samos (310–230 B C) anticipated the Copernican theory of the solar system by 1800 years. It is interesting to compare the response made to this theory in ancient Greece with that later made in Christian Europe. The contemporaries of Aristarchus scientifically discussed the theory as a possible hypothesis and finally felt it was implausible because they were unable to observe a certain consequence (stellar parallax) of the theory. The contemporaries of Copernicus condemned the theory,

---

* For example: (1) A man has 12 coins, all of which appear exactly alike, but one of which is counterfeit and does not weigh the same as a genuine coin. He has at his disposal a beam balance, but no weights. How can he detect the false coin, and whether it is light or heavy, in not more than three weighings? (2) Given 31 coins, of which exactly one is false, being of weight unequal to that of the other 30, and also given a spring balance, in five weighings determine which coin is false, the weight of the false coin, and the weight of a good coin. (3) As is well known, Lower Slobbovia is too poor a country to afford its own mint. There are N coiners engaged in making Rasbuckniks, the local currency, to government specifications. However it is suspected that some of them may be counterfeiting by introducing base metal into the alloy. Any pair of counterfeits will weigh the same, although they are slightly different from the weight of a good coin. Each coiner produces either all good coins or all counterfeits. With one guaranteed good coin, a set of infinitely refinable weights, a beam balance, and as many coins from each coiner as may be needed, determine in three weighings whether any of the coiners is dishonest, and which ones.

and persecuted its believers, because it removed man from the central position of the universe.

314° *The Greeks versus the Romans.* Referring to the death of Archimedes (see Item 77° of *In Mathematical Circles*), A. N. Whitehead, the philosopher, pointed out the difference between the Greek and the Roman mind by saying, "No Roman ever died in contemplation over a geometrical diagram."

Petr Beckmann has divided the men who make history into two classes, the thinkers and the thugs. The Greeks were thinkers and the Romans were thugs. The general law seems to be that the thugs always win, but the thinkers always outlive them.

315° *Boxing the mathematical compass.* Some years ago Walter Crosby Eells undertook the task of listing the one hundred most eminent mathematicians in the order of their eminence (see his paper, "One hundred most eminent mathematicians," *The Mathematics Teacher*, November, 1962, pp. 582–588; also see Item 254° of *In Mathematical Circles*). For his method of listing, Eells chose the so-called *space method*, wherein order of eminence is determined simply by measuring the amount of space occupied by the accounts of the various mathematicians in a selected group of standard encyclopedias and biographical dictionaries of the world. His list runs from Newton, as number one, to Zeno, as number one hundred. In the frontispiece to this volume, the reader will find, in clockwise order from north back to north again, abbreviations of the first thirty-two mathematicians on Eell's list, and the intrigued reader may care to try to figure out the names involved before peeking into Eell's article cited above.

316° *Internationalism.* The absolute calculus, as we know it today, originated in the fundamental paper, "Méthodes de calcul différentiel absolu et leurs applications," written in *French*, by the two *Italian* mathematicians G. Ricci and T. Levi-Civita, and published in the *German* journal *Mathematische Annalen*, Vol. 54 (1901)—a splendid example of the international spirit of mathematics.

317° *The performers and the creators.* All great musicians can be classified as either expert virtuosos or expert composers, and a few have been both. Similarly, all great mathematicians can be classified as either expert formal operators or expert creators of theory, and a few have been both. For example, Euler was primarily a great formal operator, Lagrange a great theorist, and Gauss was pre-eminently accomplished as both. Thus Euler was like a Heifitz, Lagrange a Beethoven, and Gauss a Johann Sebastian Bach.

318° *Virtuosity versus depth.* "The pursuit of pretty formulas and neat theorems can no doubt quickly degenerate into a silly vice, but so can the quest for austere generalities which are so very general indeed that they are incapable of application to any particular."—E. T. BELL

319° *A lesson in freedom.* In a branch of mathematics, once one has chosen a consistent set of postulates, one has complete freedom in the development of that branch of mathematics. In other words, in mathematics there is a restricted freedom—a freedom subject to some set of consistent basic laws. This is true of any creative art. The water-colorist and the oil painter have freedom, but only so far as their media permit. A composer of orchestral music has freedom, so far as the instruments he is writing for permit. A poet has freedom, though his language may be confined by a grammar and his sonnets by a form. In short, any freedom to create has certain basic confinements; the task is to use this purposefully restricted freedom to achieve something new or significant. Freedom with no accepted underlying restrictive laws is anarchy, and is hardly creative. To be free to do anything one pleases amounts to being free to do absolutely nothing.

320° *A difference between mathematics and other arts.* The following thought was expressed by Cornelius Lanczos.

Most of the arts, as painting, sculpture, and music, have emotional appeal to the general public. This is because these arts can be experienced by some one or more of our senses. Such is not true of the art of mathematics; this art can be appreciated only by mathematicians, and to become a mathematician requires a long period of intensive training. The community of mathematicians is similar to an imaginary community of musical composers whose only satisfaction is obtained by the interchange among themselves of the musical scores they compose.

321° *The grove of scholarship.* The following thought was expressed by Norbert Wiener

The value of a grove of sequoia trees does not lie in the returns that can be obtained from it today by some lumber company cutting up the trees into boards. Such returns would soon be dissipated and the grove would be forever gone. The value of the grove of sequoia trees lies in its benefit to all mankind by its mere presence among us over a very long period of time, and to cut up the grove into lumber today would be to sell out something of our future and of the futures or our children and grandchildren.

In much the same way, the value of a university of scholars cannot be counted by its present cost to a state or a government. Most of the discoveries and ideas achieved by the scholars of today have little chance of redounding to the benefit of those now paying the bill. The value of the university of scholars lies in its long-range good for all mankind, and to discontinue the university today would be to sell out something of our future and of the futures of our children and grandchildren.

Thus neither the value of the grove of sequoia trees nor the value of a tradition of scholarship can be measured by the cash-worth of the moment; rather they rest in the belief that the continued existence of the trees and the continued development of knowledge are each fine things that conspire for the good of all men.

322° *A law of genetics.* There have been so many instances in the academic world of mathematicians marrying the daughters of their professors, that a peculiar law of genetics concerning mathematical ability has been formulated: "Mathematical ability is not inherited from father to son, but from father-in-law to son-in-law."

323° *Cheating.* A student caught copying from the paper of another student during a mathematics examination defended himself by saying, "I was merely trying to create a good impression"

324° *Equivalence and identity.* Though it is true that $\sin x = O(1)$, it is not true that $O(1) = \sin x$.

"Honesty is the best policy." Therefore, if one acts so as to do the best for oneself, one is assured of acting honestly.

325° *A proposal and its acceptance.* In Ripley's *Believe It or Not, 15th Series*, a story is told of a mathematics teacher Robert Greer, who taught at the Mount School in York, England. In 1880 he proposed to a girl named Anne as follows:

"If $R = 1/2$ and $A = 1/2$,
Then $R + A = 1$,
But $R - A =$ nothing at all."

He received the happy answer: "Let it be $R + A$."

326° *On large numbers.* The relative sizes of a million, a billion, and a trillion can be made striking by pointing out that a million seconds is equivalent to about 12 days, a billion seconds to about 32 years, and a trillion seconds to about 32 millennia.

327° *Dependence.* In an address entitled "The function concept in elementary teaching and in advanced mathematics," delivered in Albuquerque, New Mexico, on April 26, 1938, before the Southwestern Section of the Mathematical Association of America and the Mathematics Section of the Southwestern Division of the American Association for the Advancement of Science, E. R. Hedrick made the following interesting remarks:

> That the bare fact of dependence requires intelligent thought may be illustrated best by an example. For many centuries, the engineers of the Roman Empire constructed aqueducts, and they supplied water to cities for domestic purposes and to farmers for irrigation. The quantities of water furnished were determined solely by the size of the opening through which the water was delivered. Apparently the first man to notice that the amount of water that would pass through a given opening depends also on the *pressure* (or "head") was Leonardo da Vinci. He also stated the approximate formula that is still in use by engineers. In this country, even during the latter half of the nineteenth century, water rights for irrigation were stated in "inches," that is, per square inch of opening, without regard to the pressure; many legal battles have resulted, especially in California, as a result of this practice.

I have stated this as a conspicuous instance of the failure, even by fairly intelligent persons vitally concerned, to note the dependence of one quantity upon another.

328° *The Alfonsine Tables.* The first major trigonometric work in Western Europe was the revision of the Ptolemaic planetary tables. This work was completed about 1254 at Toledo, under the reign of the Spanish ruler Alfonso X. The revised tables, which were highly esteemed by later astronomers, became known as *the Alfonsine Tables*, thus glorifying in the history of mathematics the name of a deserving patron of learning.

329° *Bad news.* The mathematics teacher announced a coming test to his class by first writing the word "gnu" on the blackboard and then saying, "As Mrs. Gnu said to Mr. Gnu when he visited her in the maternity ward, 'I have gnus for you.'" He then stated that at the next class meeting there would be a mathematics test on the first two chapters of the textbook.

330° *A consuming proof.* The theorem and proof of Figure 66 were seen on the blackboard of a mathematics classroom.

ROGER HOUGH

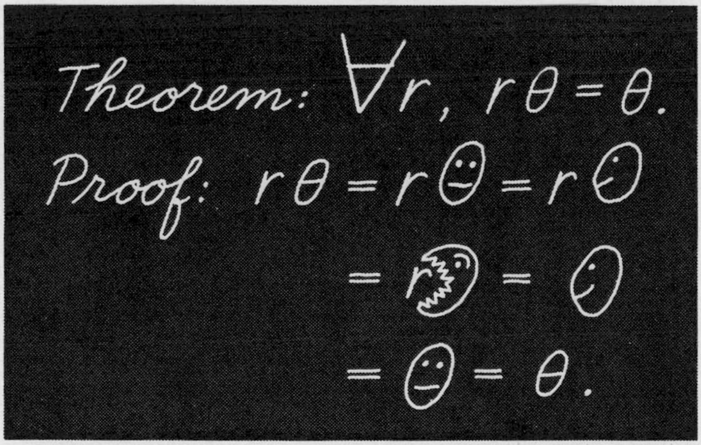

FIGURE 66

331° *What a sprouting acorn said.* "Gee, I'm a tree." (Geometry.)

332° *A revolutionary sputnik.* On October 4, 1957, the Russians launched the first artificial satellite, Sputnik I, which, until it fell to the earth on January 4, 1958, traveled at a speed of 18,000 miles per hour and circled the earth once about every ninety-five minutes. Russia and the United States have since launched other artificial satellites, some bearing passengers, as a monkey, a pair of dogs, and human beings. It is unofficially rumored that Russia is now contemplating putting into orbit around the earth a sputnik carrying two cows. Apparently this is being considered by the Russians so that they might, in a sense, outdo the Americans, for should the feat be accomplished it will be known as "the herd shot round the world."—ROBERT A. ESTES

333° *The queen of the sciences.* Professor L. M. Passano, of our Department of Mathematics at M.I.T., was also a literary man. He was opposed, as a humanist, to the prominence the department was giving to applied mathematics. As he wrote in a memorandum, "This trend will make the queen of the sciences into the quean of the sciences."— DIRK J. STRUIK

334° *An accidental immortality.* When Ramanujan was sixteen, he happened upon a copy of Carr's *Synopsis of Mathematics.* This chance encounter secured immortality for the book, for it was this book that suddenly woke Ramanujan into full mathematical activity and supplied him essentially with his complete mathematical equipment in analysis and number theory. The book also gave Ramanujan his general direction as a dealer in formulas, and it furnished Ramanujan the germs of many of his deepest developments.

335° *The Scopes trial.* The facts of the famous Scopes trial of Tennessee are usually presented somewhat as follows:

In March of 1925, the Tennessee legislature passed a law forbidding the teaching in any of the public schools of the state of "any theory which denies the story of the Divine creation of man as taught in the Bible." The law was obviously aimed at the spreading popular theory in educational circles of the Darwinian theory of evolution.

A few months later, John T. Scopes, a biology teacher in the high school of Dayton, Tennessee, was indicted for teaching evolution, contrary to the recently passed state law, and was brought to trial on July 10th. The prosecution was conducted by the state attorney general with the oratorical assistance of William Jennings Bryan, an ardent fundamentalist in religion. The defence was led by the famous lawyer Clarence W. Darrow, assisted by Arthur G. Hays. After a spectacular eleven-day trial that roused nationwide ridicule, Scopes was found guilty and fined $100. Five days later, while resting in Dayton after the trial, Bryan suddenly died. Two years later, in 1927, the state appellate court reviewed the case and reversed the decision, thus preventing the further embarrassment to the state of an appeal to the United States Supreme Court.

Now the above facts are neither quite complete nor quite accurate. In the first place, Scopes was a mathematics and physical education teacher at the Dayton high school, and never taught biology a single day of his life. In the second place, Scopes volunteered to stand trial in order to furnish the board of education a test case of the ridiculous state law. When Scopes died, in October 1970, these additional and generally unknown facts were publicized on the NBC Nightly News television broadcast.

336° *Hermannus Contractus.* The Swabian Count Wolverad had a son Hermann (1013–1054) who, at birth or as a young child, was so paralyzed that he could not move his lips, use his crippled hands, or manage his twisted legs. Yet this unfortunate child in time became a celebrated astronomer, mathematician, clockmaker, and creator of fine musical instruments. Because of his painfully contracted limbs, he bore the name of Hermannus Contractus, or Hermann the Lame. He received an education in the monastic school of Reichenau, later joined the Benedictine order, gave lectures in mathematics that drew about him a large group of pupils, and wrote notable works on the astrolabe, the abacus, and the number game of rithmomachia.

PRINTERS AND BOOKS

BEFORE we complete our circuit of stories we ought, perhaps, to tell some that are connected with the printing of mathematics articles and books. Here, then, is a small selection. The last few stories in the group concern jokers in mathematics books. Stories of this genre are often either unkind or indelicate. The interested reader can find another example in a generalization of the concept of osculation on the last page of Dirk J. Struik's *A Source Book in Mathematics*, 1200–1800, Harvard University Press, 1969. Dr. Struik says that the story is his own little joker; in accordance with a long tradition he has left it untranslated.

337° *The copy editor.* Octave Levenspiel once told of an author (perhaps it was himself) who used three characters called Lurch, Reel, and Stagger in the problem sets of an elementary mathematics text. The publishing house copy editor, when checking the index of the work, looked the three characters up in the text and found they were "people." Since, by adopted style, the index recorded the last names *and initials* of all persons referred to, the copy editor, in a note to the author, in connection with Lurch, Reel, and Stagger, asked, "Furnish initials, please."

338° *A typist's error.* "There should be a more vigorous treatment of rectors."

339° *A questionable practice.* There is a literary convention that numbers less than 10 should be spelled out as words. This is sometimes unsuitable in mathematics. Thus J. E. Littlewood has cited the following horrible example that he once came across: "functions never taking the values nought or one."

340° *Linguistic oddities.* A linguist would be disturbed to learn that "set E is closed" does not mean "set E is open," and that "E is dense in E" does not have the same meaning as "E is dense in itself." On the other hand, the phrases "there is more than one" and "there are fewer than two" can sound queer to a mathematician.

341° *Dash it all.* Littlewood said that he had occasion to read aloud the phrase: "where E' is any dashed [*i.e.*, derived] set." He found it necessary to place the stress with some care. (It should be realized that the British read E', not as "E prime," but as "E dash," and that to them the word "dashed" is equivalent to the popular American "damned," used as a mild imprecation.)

342° *Proofreading.* Proofreading is a boring, tedious, and onerous task, and the proofreader is often chagrined later to find one or more errors that escaped him. Professor G. H. Hardy was a meticulous mathematician and perhaps as good a proofreader of mathematical papers as could be found. He and J. E. Littlewood formed one of the most famous collaboration teams of modern times; they collaborated for thirty-five years, turning out a large number of brilliant papers. Littlewood once challenged Hardy to find a misprint on a certain page of one of their joint papers. Hardy failed; the misprint occurred in his own name: "G, H. Hardy."

343° *The compositor.* Amusing incidents occur in connection with compositors and printers of mathematical articles. J. E. Littlewood, about 1917, wrote a memorandum for the British Ballistic Office, ending with the sentence, "This σ should be made as small as possible." When the draft was printed, this sentence did not appear. Reading the memorandum, someone asked, "What is that?" Upon careful scrutiny, a little speck in the blank space at the end of the memorandum proved to be the tiniest σ imaginable; the compositor must have scoured London to secure it.

One wonders what the compositor might have done with a phrase like: "where big X is very small."

344° *A ghost-author.* In the article, "Noether's canonical curves," *Mathematical Gazette,* XIV, 23, by W. P. Milne, there is a note by the author that refers to "a paper by Guyaelf (*Proc. London Math. Soc.,* ser. 2, 21, part 5)." Guyaelf was a pseudonym used by W. P. Milne.

345° *An unkind textbook reference.* In B. M. de Kerékjártó's *Vorlesungen über Topologie,* the index has a reference to Bessel-Hagen. On the corresponding page Bessel-Hagen is not mentioned, but there is a figure (No. 27) of a face-like thing with two huge ears. Bessel-Hagen was a quiet, warm-hearted fellow, and in the rather cruel atmosphere of Göttingen was an easy prey for jokes. (See Item 314° in *Mathematical Circles Revisited*.)—DIRK J. STRUIK

346° *Another unkind textbook reference.*   In E. T. Bell's *Development of Mathematics* (second edition), the index has a reference to G. A. Miller on p. 445. You do not find the name of Miller there, only a remark on "industrious laborers little advanced beyond mathematical illiteracy."—DIRK J. STRUIK

347° *Weitzenböck's "Invariantentheorie."*   R. Weitzenböck, a professor in America, had been an officer in the Austrian army and remained a *Franzosenfresser*. If you take the initial capitals of every sentence in the introduction to his *Invariantentheorie*, you read, "Neider mit den Franzosen."—DIRK J. STRUIK

PSYCHOLOGY

IN 1902 and 1904, the Swiss periodical *L'Enseignement Mathématique* (a journal devoted to the teaching of mathematics), undertook an inquiry into the working methods and habits of mathematicians. A questionnaire containing upwards of thirty questions was mailed to a number of mathematicians, and in time replies from over a hundred of the recipients were received. The answers, along with an analysis of suggested traits and trends, were published in a 137-page memoir in 1912. This work is of cardinal interest to anyone concerned with the psychological make-up of mathematicians. More recently (1945) there has appeared Jacques Hadamard's splendid 136-page essay on *The Psychology of Invention in the Mathematical Field*.

348° *Gauss's experience.*   Carl Friedrich Gauss once referred as follows to a theorem which for years he had unsuccessfully tried to prove: "Finally, two days ago, I succeeded—not on account of my hard efforts, but by the grace of the Lord. Like a sudden flash of lightning, the riddle was solved. I am unable to say what was the conducting thread that connected what I previously knew with what made my success possible."

On another occasion, after one of his insightful flashes of intuition, Gauss exclaimed, "I have the result, but I do not yet know how to get it."

(See also Item 329° of *In Mathematical Circles*.)

349° *Hamilton's experience.* In Item 338° of *In Mathematical Circles*, it was told how William Rowan Hamilton, after wrestling with the problem of quaternion multiplication for fifteen years, was suddenly struck with the unorthodox idea of abandoning the traditional commutative law of multiplication. The event occurred in 1843, when Hamilton was thirty-eight years old and had already for two years been President of the Royal Irish Academy. It is interesting to read Hamilton's own description of his discovery as given in a letter to his son:

On the 16th day of October, which happened to be a Monday, and Council Day of the Royal Irish Academy, I was walking in to attend and preside, and your mother was walking with me along the Royal Canal, to which she had perhaps driven; and although she talked with me now and then, yet an undercurrent of thought was going on in my mind, which gave at last a result, whereof it is not too much to say that I felt at once the importance. An electric circuit seemed to close; and a spark flashed forth, the herald (as I foresaw immediately) of many long years to come of definitely directed thought and work, by myself if spared, and at all events on the part of others, if I should even be allowed to live long enough distinctly to communicate the discovery. Nor could I resist the impulse—unphilosophical as it may have been— to cut with a knife on a stone of Brougham Bridge, as we passed it, the fundamental formula with the symbols i, j, k: namely

$$i^2 = j^2 = k^2 = ijk = -1,$$

which contains the solution of the problem, but of course as an inscription has long since mouldered away. A more durable notice remains, however, in the Council Book of the Academy for that day, which records the fact that I then asked for and obtained leave to read a paper on Quaternions, at the first general meeting of the session, which reading took place accordingly on Monday the 13th of November following.

350° *Poincaré's experience.* The best known and most quoted example of the "preparation-incubation-illumination" sequence in mathematical discovery is that told by Henri Poincaré:

For fifteen days I struggled to prove that no functions analogous to those I have since called *Fuchsian functions* could exist; I was then very ignorant. Every day I sat down at my work table where I spent an hour or two; I tried a great number of combinations and arrived at no result. One evening, contrary to my custom, I took black coffee; I could not go to sleep; ideas swarmed up in clouds; I sensed them clashing until, to put it so, a pair would hook together to form a stable combination. By morning I had established the existence of a class of Fuchsian functions, those derived from the hypergeometric series. I had only to write up the results, which took me a few hours.

Next I wished to represent these functions by the quotient of two series; this idea was perfectly conscious and thought-out; analogy with elliptic functions guided me. I asked myself what must be the properties of these series if they existed, and without difficulty I constructed the series which I called thetafuchsian.

I then left Caen, where I was living at the time, to participate in a geological trip sponsored by the School of Mines. The exigencies of travel made me forget my mathematical labors; reaching Coutances we took a bus for some excursion or other. The instant I put my foot on the step the idea came to me, apparently with nothing whatever in my previous thoughts having prepared me for it, that the transformations which I had used to define Fuchsian functions were identical with those of non-Euclidean geometry. I did not make the verification; I should not have had the time, because once in the bus I resumed an interrupted conversation; but I felt an instant and complete certainty. On returning to Caen I verified the result at my leisure to satisfy my conscience.

I then undertook the study of certain arithmetical questions without much apparent success and without suspecting that such matters could have the slightest connection with my previous studies. Disgusted at my lack of success, I went to spend a few days at the seaside and thought of something else. One day, while walking along the cliffs, the idea came to me, again with the same characteristics of brevity, suddenness, and immediate certainty, that the transformations of indefinite ternary quadratic forms were identical with those of non-Euclidean geometry.

On returning to Caen, I reflected on this result and deduced its consequences; the example of quadratic forms showed me that there were Fuschian groups other than those corresponding to the hypergeometric series; I saw that I could apply to them the theory of thetafuchsian functions, and hence that there existed thetafuchsian functions other than those derived from the hypergeometric series, the only ones

I had known till then. Naturally I set myself the task of constructing all these functions. I conducted a systematic siege and, one after another, carried all the outworks; there was however one which still held out and whose fall would bring about that of the whole position. But all my efforts served only to make me better acquainted with the difficulty, which in itself was something. All this work was perfectly conscious.

At this point I left for Mont-Valérien, where I was to discharge my military service. I had therefore very different preoccupations. One day, while crossing the boulevard, the solution of the difficulty which had stopped me appeared to me all of a sudden. I did not seek to go into it immediately, and it was only after my service that I resumed the question. I had all the elements, and had only to assemble and order them. So I wrote out my definitive memoir at one stroke and with no difficulty.

351° *Hadamard's experience.*   A phenomenon, closely related to dreaming, reported by many mathematicians is the immediate illumination of a solution to a problem at the very moment of sudden awakening. The French mathematician Jacques Hadamard (1865–1963) has related such an experience that happened to him. Some noise, it seems, suddenly awakened him, and immediately a long-searched-for solution to a problem appeared to him without any reflection whatever on his part, and the solution was in a direction entirely different from any he had previously tried.

352° *Loss of interest.*   A number of mathematicians have confessed to a loss of interest in a problem or investigation immediately after a solution has been obtained. Unfortunately, this "letdown" usually coincides with the time when the solution should be recorded. Apparently the excitement and interest lie in the pursuit of the solution, not in the writing up of the result. Jakob Steiner (1796–1863) frequently discovered geometrical properties with such speed that he disliked the tedium of writing up their proofs, and so he left many remarkable discoveries as puzzles for his successors to establish. (See also Item 203°.)

353° *A mathematical dream.*   Very few mathematical dreams have been described. A remarkable instance, however, was reported by the American mathematician Leonard Eugene Dickson (1874–

1954). His mother and her sister were rivals in geometry at school, and once they spent an entire evening in vain over a certain problem. That night, Dickson's mother dreamed of the problem and while asleep developed a solution in a voice so clear that her sister, who awoke, was able to take notes. The next morning in geometry class, the sister presented the solution (which was correct), while Dickson's mother could not recall it.

One is reminded of Marie Agnesi's mathematical accomplishments while sleepwalking (see Item 273° of *In Mathematical Circles*).

354° *The mathematics bump.* The German physician and phrenologist, Franz Joseph Gall (1758–1828), put forth the theory that any particular mental aptitude was connected with a greater development of some special part of the brain, and that this greater development of some part of the brain resulted in a bulge in the corresponding part of the skull. Thus a person's mental characteristics could be determined by examining the shape of his head. All creative mathematicians, for example, would possess a "mathematical bump" at some well-located place of the skull. Modern researches by anatomists and neurologists have proved that Gall's phrenology is physiologically unsound.

355° *Servants, not masters.* There is a distinction between discovery and invention. The former concerns something that already existed but had not yet been perceived; the latter concerns something that never existed until it was perceived. Thus on the one hand Columbus discovered America; America was already in existence before Columbus found it. On the other hand, Edison invented the light bulb; there was no light bulb until Edison created one.

We now see a great difference between art and mathematics. The works of an artist are inventions; his works did not exist before he created them. The findings of a mathematician are discoveries; his findings already existed before he came upon them. A mathematical truth, though not yet known to us, pre-exists, and thus imposes on the mathematician certain paths that he must follow if he is to find it. This is the root of a remark made by Charles Hermite (1822–1901): "We are servants rather than masters in mathematics." In contrast, artists are masters rather than servants.

356° *At what age?* The first question on the questionnaire distributed by *L'Enseignement Mathématique* was, in part, "At what time, as well as you can remember, . . . did you begin to be interested in mathematics?" There were ninety-three replies to this question: thirty-five said it was before the age of ten; forty-three said it was from eleven to fifteen; eleven said it was from sixteen to eighteen; three said it was from nineteen to twenty; and one said it was at twenty-six.

357° *Morning or evening?* Question 27 on the questionnaire distributed by *L'Enseignement Mathématique* was, "Would you rather work [at mathematics] in the morning or in the evening?" Answers to this question varied but indicated an interesting and possibly significant trend. Though there were exceptions, the mathematicians of the northern races said they preferred to work at night, while the mathematicians of the Latin races favored the mornings. It was also brought out that among the evening-workers, the late concentration often induced insomnia in later age, with the result that work hours were reluctantly changed from evening to morning.

358° *Preparation-incubation-illumination.* The three-part sequence of mathematical discovery—preparation, incubation, illumination—has its counterparts in everyday life. How often it happens that the name of some person or place elludes our most strenuous efforts at recollection, and then suddenly occurs to us when we are no longer thinking of it. And how well-known is the saying, "Sleep on it." Professor Kurt O. Friedrichs has said, "Creative ideas come mostly of a sudden, frequently after great mental exertion, in a state of mental fatigue combined with physical relaxation."

359° *Two kinds of discovery.* There seem to be two kinds of discovery. In one kind, the goal is given first and then the mind goes from the goal to the means, that is, from the question to the solution. In the other kind, the mind goes from the means to the goal, that is, the mind first discovers a fact and then seeks a use for it. In mathematics, and elsewhere, most significant discoveries are of the second kind. As Hadamard has put it, "Practical application is found by not looking for it, and one can say that the whole progress of civilization rests on that principle." An outstanding example in mathematics is the exhaustive study of the conics by the Greeks, and then, some two thousand years later, Kepler's stunning application of the Greek findings to the movement of the planets in the solar system. The physicist and artist Duhem once compared Hadamard to a landscape painter who in his studio creates a landscape painting and then leaves the studio to find in nature some landscape fitting his painting.

360° *The thought curve.*

"Very good. Now, Mr. Soames, we will take a walk in the quadrangle, if you please."
Three yellow squares of light shone above us in the gathering gloom.
"Your three birds are all in their nests," said Holmes, looking up. "Halloa! What's that? One of them seems restless enough."
It was the Indian, whose dark silhouette appeared suddenly upon his blind. He was pacing swiftly up and down his room.
....
". . . Why should he be pacing his room all the time?"
"There is nothing in that. Many men do it when they are trying to learn anything by heart."

The Adventures of the Three Students—A. CONAN DOYLE

Like the Indian in *The Adventure of the Three Students*, many scholars find that concentration and the thinking process are stimulated by regular and rhythmic pacing—that is, that "legs are the wheels of thought." This has certainly been true of many of the great creative mathematicians (see, for example, Items 167° and 234°).

Clearly the most convenient place for this kind of pacing is in the quiet and privacy of one's own study. But if one is to pace about in an ordinary study room, the curve or path of pacing clearly must satisfy some very definite requirements. Let us examine some of these requirements of the *thought curve*.

1. It would be convenient to have the curve fit neatly into a rectangle. This follows from the fact that the largest clear space amongst the furniture pieces is usually either rectangular or readily converted to a rectangle.

2. The curve should be closed. The convenience of this is obvious.

3. The curve should be planar. This is by no means necessary,

but is desirable on at least two counts, namely the difficulty of setting up a nonplanar path and the fact that for regular and rhythmic pacing one would not care to be climbing at some point of the path and then hurriedly descending at another.

4. The curve should be continuous. Again, within limits, this is not necessary, but is certainly convenient. For, roughly speaking, there are two major types of discontinuity.

a. Discontinuity caused by an infinite branch of the curve. The dimensions of the room and the continuous nature of time exclude the possibility of having this type of discontinuity.

b. Discontinuity caused by a break in the curve. If there is such a break, it must, for humanly physical reasons, be less than one's maximum leap. But further, since a sudden leap is contrary to the notion of regular and rhythmic pacing, we must actually have any such break in the curve less than the length of one's ordinary step. Now, since the curve is closed, it is unlikely, unless great care is taken in laying out its length, that after one complete cycle the pacer will arrive at the break just right to clear it by his next step. For this reason, the longer one walks the curve, the smaller should be any such break. The best thing is to settle for continuity.

5. The curve should possess a continuous first derivative (except for some possible infinite values). Regularity and rhythm in pacing preclude the idea of a sudden or sharp turning. One's turning must be smooth, whence the curve must have a continuous first derivative as prescribed.

6. The curve should have a curvature never greater than, say, that of a circle of $2\frac{1}{2}$ feet. For, clearly, any greater curvature tends to become too sharp for comfort in regular and rhythmic pacing.

7. The curve must contain at least two inflections. This follows because the curvature cannot persist in one direction without leading to dizziness. But, of course, a closed curve with opposite curvatures must contain an even number of inflections, whence the requirement.

8. Symmetry in the curve is helpful, as symmetry supports regular and rhythmic pacing.

Now we are very fortunate to find in mathematics a curve that satisfies all the above requirements—a curve which is indeed ideal for thought stimulation. So at this point let us offer the lemniscate of

Jakob Bernoulli as the perfect *thought curve*.* This curve appears as in Figure 67 and has for a polar equation

$$r^2 = a^2 \cos 2\theta,$$

and for a Cartesian equation

$$(x^2 + y^2)^2 = a^2(x^2 - y^2).$$

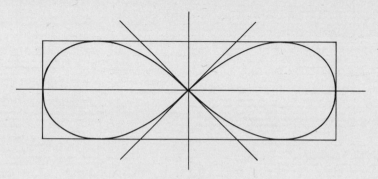

FIGURE 67

---

* Royal-Dawson, of England, has recommended the lemniscate of Bernoulli as a transition spiral in highway designing. Thus this lemniscate is not only a curve conducive for thought, but also a curve comfortable for driving.

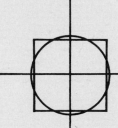

ADDENDA

From esthetic appreciation
to some associated quotes

ESTHETICS

We have, at this point of our three works (*In Mathematical Circles, Mathematical Circles Revisited, Mathematical Circles Squared*), told a total of 3(360) = 1080 mathematical stories and anecdotes. We now add nine more items, all dealing with the esthetic aspect of mathematics, to raise the total to 1089 = $(33)^2$, the first square number above 1080, so that we may complete *Mathematical Circles Squared*.

1081° *The esthetic appreciation of mathematics.* [The two following items are adapted, with permission, from H. E. Huntley's attractive little book, *The Divine Proportion, a Study in Mathematical Beauty*, Dover Publications, Inc., 1970.]

The feeling for beauty in mathematics is infectious. It is caught, not taught. It affects those with a flair for the subject. I well remember when it happened to me, as a very young undergraduate of Bristol University. It was a seminal experience in life.

The late Peter Frazer, Lecturer in Mathematics, a lovable man and a brilliant teacher, was discussing cross ratios. Swiftly he chalked on the blackboard a fan of four straight lines, crossed them with a transversal and wrote a short equation; he stepped down from the dais and surveyed the figure. I cannot of course recall precisely what he said but it went something like this. Striding rapidly up and down between the class and the blackboard, waving his arms about excitedly, with his tattered gown, green with age, billowing out behind him, he spoke in staccato phrases. "Och, a truly beautiful theorem! Beautiful! . . . Beautiful! Look at it! *Look at it*! What simplicity! What economy! Just four lines and one transversal." His voice rises in a crescendo. "What elegance! *Any* lines, *any* transversal! Its generality is *astonishing*." Then, muttering to himself, "Beautiful! . . . beautiful! . . .," he stopped, slightly embarrassed (he was from Aberdeen), and returned to earth.

The students were amused. But not all. Sparks from that blazing enthusiasm fell on at least one boy. He took fire and that fire was never extinguished.—H. E. HUNTLEY.

1082° *Esthetic appreciation smothered.* On an occasion when I was speaking to a Student Christian Movement meeting of sixth formers I happened to remark, incidentally, that the famous theorem of Pythagoras was "a thing of beauty." The explosion of derisive laughter that greeted this innocent remark was shattering. The reason for the outburst was, in my view, very simple. Everyone knew that what I had said was true, but to admit it involved "wearing one's heart on one's sleeve," and this "isn't done" by sixth formers. One seldom hears the adjective "beautiful" from the lips of an adolescent; his private feelings are not for public display.—H. E. HUNTLEY

1083° *The practical versus the esthetical.* Practicality is an important feature of mathematics, but what is practically useful in mathematics in one era may no longer be practically useful in another, and there are large parts of mathematics for which little practical use has been found. We find the same situation in science. Einstein claimed that, though the world can be understood in terms of reason, the criteria for the acceptance of a theory are, in the last analysis, esthetical (see Item 313° of *In Mathematical Circles*). The same thesis was maintained by Dirac when he said, "It is more important to have beauty in one's equations than to have them fit experiment."

It follows that mathematics, and perhaps science too, should be studied more with a view to its esthetic values than with a utilitarian objective in mind.

1084° *Popular appeal of mathematics.* There is a widespread belief that appreciation of the esthetic qualities of mathematics is rare and that very few people are sensitive to the kind of beauty found in mathematics. The universal popularity of games with a mathematical basis, like *Go* in Japan and chess in Russia, seems to contradict this view. There is even a vast literature, in both periodical and book form, devoted to the intricacies of some of these games, such as chess and checkers. The very adjectives used by mathematicians in describing a particularly fine proof or theorem are frequently employed by the players of games. Thus an end game in chess may be described as "beautiful," a certain checkmate as "neat," and the solution of some chess problems as "elegant." As further supporting evidence for the really widespread appeal of mathematical reasoning there is the great popularity of mathematical puzzles found in the puzzle columns of newspapers and magazines, and in the fact that books on popular mathematics are sold yearly as paperbacks by the millions. If mathematics in the lower schools should be taught with a stress on its esthetical charms, the subject would win ever so many more converts.

1085° *The guide for the choice of research subjects.* Since it very seldom happens that significant mathematical research is undertaken directly with a view toward practical application (see Item 359°), the question naturally arises: How are mathematicians to select their subjects for research? This choice is certainly important, for on it we form our judgment of the mathematical stature of the researcher. There seems no doubt as to the proper basis for the choice of topics— it is an esthetic sensibility, a finely developed feeling for what is beautiful. Just as there is a literary taste and an artistic taste, there is this mathematical taste, and some possess it more deeply or with more refinement than others, and these latter tend to develop more lasting and more significant mathematics. The applications almost always appear in time.

The Greek development of the conic sections proved to be lasting and significant mathematics, not because it was applicable (the applications of the conics came centuries later), but because that development was intrinsically beautiful. The Greeks were such remarkable mathematicians because they somehow or other could distinguish the more beautiful from the less beautiful. The Hindus generally lacked this ability, and their mathematics has proven to be of very uneven quality, good and bad appearing side by side. The Greeks had the ability to select the good and discard the bad—the good being the esthetically pleasing and the bad not.

1086° *Judging a graduate research student.* One of the most disappointing things is to have one of your graduate students come up to you and ask for possible subjects for research. One's almost immediate reaction is to classify such a student as probably at best second rate. If a student, having attended college courses in mathematics for three or four years, has not yet perceived that there are attractive matters still awaiting deeper development, then that student appears not to possess the esthetic taste mentioned in the previous item and seems predoomed to mediocrity in his field.

1087° *A case history.* Hadamard wrote his doctoral dissertation under the direction of Hermite. When Hadamard presented his thesis for examination, Hermite noted that it would be nice if Hadamard could find an application. Now, at the time, Hadamard had no application and had not even thought about finding one, but between the time his thesis was handed in and the day it was accepted, he became aware of an important question whose answer was supplied by the thesis. Thus his paper did turn out to be *useful*, showing that his sheer feeling for the attractiveness of the problem of his thesis had guided him in the right direction.

A similar event occured to Hadamard in connection with his beautiful theorem of 1893 on determinants. Seven years later, in 1900, the Fredholm theory found Hadamard's determinant theorem to be essential.

1088° *Another case history.* In the eighteenth century, Jean Bernoulli became interested in the curve, connecting two points not in the same horizontal plane, down which a weighted particle will slide in the least time. Now it was the sheer beauty, and the entirely different nature, of this problem that attracted Bernoulli. That problems of this new short (that is, consequences of the "calculus of variations") would be vital to the development of mechanics at the turn of the century was not, indeed could not have been, suspected by Bernoulli at the time.

1089° $= [(33)^2]°$ *Ten pertinent quotations.* The following ten quotations support the contention that a sense of beauty is the principal drive of discovery in mathematics.

1. The mathematician's patterns, like the painter's or the poet's, must be *beautiful*; the ideas, like the colors or the words, must fit together in a harmonious way. Beauty is the first test: there is no permanent place in the world for ugly mathematics.—G. H. HARDY

2. It is, so to speak, a scientific tact, which must guide mathe-

maticians in their investigations, and guard them from spending their forces on scientifically worthless problems and abstruse realms, a tact which is closely related to *esthetic tact* and which is the only thing in our science which cannot be taught or acquired, and is yet the indispensable endowment of every mathematician.—H. HANKEL

3. The mathematician requires tact and good taste at every step of his work, and he has to learn to trust to his own instinct to distinguish between what is really worthy of his efforts and what is not.—J. W. L. GLAISHER

4. It was not alone the striving for universal culture which attracted the great masters of the Renaissance, such as Brunellesco, Leonardo da Vinci, Raphael, Michael Angelo, and especially Albrecht Dürer, with irresistable power to the mathematical sciences. They were conscious that, with all the freedom of the individual fantasy, art is subject to necessary laws, and conversely, with all its rigor of logical structure, mathematics follows esthetic laws.—F. RUDIO

5. Mathematics, rightly viewed, possesses not only truth, but supreme beauty—a beauty cold and austere, like that of sculpture, without appeal to any part of our weaker nature, without the gorgeous trappings of painting or music, yet sublimely pure, and capable of a stern perfection such as only the greatest art can show.—B. RUSSELL

6. The true mathematician is always a good deal of an artist, an architect, yes, of a poet.—A. PRINGSHEIM

7. Who has studied the works of such men as Euler, Lagrange, Cauchy, Riemann, Sophus Lie, and Weierstrass, can doubt that a great mathematician is a great artist?—E. W. HOBSON

8. Mathematics has beauties of its own—a symmetry and proportion in its results, a lack of superfluity, an exact adaptation of means to ends, which is exceedingly remarkable and to be found only in the works of the greatest beauty. . . . When this subject is properly . . . presented, the mental emotion should be that of enjoyment of beauty, not that of repulsion from the ugly and the unpleasant.

J. W. A. YOUNG

9. A peculiar beauty reigns in the realm of mathematics, a beauty which resembles not so much the beauty of art as the beauty of nature and which affects the reflective mind, which has acquired an appreciation of it, very much like the latter.—E. E. KUMMER

10. He must be a "practical" man who can see no poetry in mathematics. — W. F. WHITE

INDEX

References are to items, *not* to pages. A number followed by the letter *p* refers to the introductory material just preceding the item of the given number (thus 181*p* refers to the introductory material immediately preceding Item 181°).

ABACISTS, 17

Abacus (*See* Counting board)

Abel, N., 181*p*, 181, 182, 183

Memoir on a general property of a very extensive class of transcendental functions, 181

Absolute calculus, 316

Account of the Rev. John Flamsteed, the First Astronomer-Royal, An (F. Baily), 198

Ackermann, W., 252

Adare, Lord, 137

Adventure of the Three Students, The (A. Conan Doyle), 360

Agnesi, M. G., 353

Airy, Sir G. B., 138

Alfonsine tables, 328

Alfonso X, 328

Algebra, 160, 325

Algorists, 17

American Association for the Advancement of Science, 327

American Fur Company, 196

American Mathematical Monthly, The, 91*p*, 109, 110, 111, 112, 113

Analytic geometry, 81

Apollodorus, 77

Archimedes, 314

Aristarchus, 313

Attic numerals, 12

Ausdehnungslehre (H. G. Grassmann), 297

Autobiography (B. Russell), 86

Automated data processing system, 26

BABYLONIAN cuneiform numerals, 64

Bach, J. S., 317

Bailey, F. L., 186

Bailey's *English Dictionary,* 15

Baily, F., 198

An Account of the Rev. John Flamsteed, the First Astronomer-Royal, 198

Barrie, J. M., 144*p*, 144, 145, 146, 151, 152, 153, 156

An Edinburgh Eleven, Pencil Portraits from College Life, 144*p*

Bayly, Miss, 138

Beckmann, P., 314

Beethoven, L. von, 317

Believe It or Not, 15th series (R. Ripley), 325

Bell, E. T., 138, 318

The Development of Mathematics, 346

Bergholt, E., 43

Berlin Academy of Sciences, 199

Bernoulli, Jakob, 308, 360

motto, 308

Bernoulli, Jean, 1088

Bernstein, F., 228

Bernstein, S., 228

Besicovitch, A. S., 311

Bessel, F. W., 205

Bessel-Hagen, 345

Bieberbach, L., 265, 266

Binary system, 11

Binet intelligence tests, 168

Birkhoff, G. D., 173, 264

Blaschke, W., 296*p*, 304

Blissard, J., 114*p*, 114, 115

Blumenthal, O., 281, 282

Bohr, H., 289, 290

Bohr, N., 289

Bolyai, W., 202

Bonaparte, N., 174*p*, 174, 175

Book of Permutations, 10, 31

Born, M., 249, 270, 287, 288
Bose, R. C., 46
Bouligand, G., 100
Bourbaki, N., 174*p*, 179
Bowditch, N., 226
Bragdon, C. F., 49
   More Lives Than One, 49
Briggs, H., 172
British Admiralty, 198
British Association, 117, 136, 141
British Ballistic Office, 343
Brooke, M., 163
Brooks, R. C., 62
Brougham, Lord, 117
Brouwer, L. E. J., 259, 266
Brunellesco (Filippo Brunelleschi) 1089
Brunswick, Duke and Duchess of, 189
Bryan, W. J., 335
Budget of Paradox, A (A. De Morgan), 118, 198
Bulletin de la Société de Mathématique de France, 178

Calculus, 360
Calculus of variations, 1088
Cambridge, Duke of, 200
Cambridge University, 116, 144
Carathéodory, C., 259, 280
Carpenter's *Mental Physiology*, 193
Carr's *Synopsis of Mathematics*, 334
Cauchy, A. L., 181, 1089
Cayley, A., 30*p*, 167, 206
Chinese numerals, 64
Chrystal, G., 144*p*, 151, 152, 153, 154, 155, 156, 157
Clebsch, A., 226
Clifford, W. K., 114*p*, 122, 123, 124, 125, 187
Code Napoléon, 175
Colaw, J. M., 109
Coleridge, S. T., 135, 212
Columbia University, 91*p*, 92
Columbus, C., 355
Compositio Mathematica, 113*n*
Computers, 26, 27, 28, 46
  MANIAC, 72
Copernican theory, 313
Copernicus, N., 187, 313
Corey, S. A., 111
Cotes, R., 114*p*, 125
Counting board, 13, 14, 17
Courant, R., 264, 265, 307
Course in Pure Mathematics, A (G. H. Hardy), 129
Cromwell, O., 14
Cross ratio, 1081
Cupid's problem, 113
Cybernetics (N. Wiener), 105

Darboux, J. G., 184, 262
Darrow, C. W., 335
Darwinian theory of evolution, 335
Da Vinci, L., 77, 305, 1089
Declaration to the Cultured World, 263
Dedekind cut, 129
Definite integrals, 119
Defoe, D., 118
Dehn, M., 283
Delphian oracle, 82
De Morgan, A., 114*p*, 116, 117, 118
   A Budget of Paradoxes, 118, 198
De Morgan, S. E. (Mrs. A. De Morgan), 118
Descartes, R., 21, 201
Development of Mathematics, The (E. T. Bell), 346
Diabolic magic squares, 30
Dickson, L. E., 353
Differential geometry, 207
DIN (Deutsche Industrie Normen), 85
Dirac, P. A. M., 1083
Dirichlet, P. G. Lejeune, 202
Discovery, 359
Discovery versus invention, 355
Disquisitiones arithmeticae (C. F. Gauss), 202
Dissection(s)
   equilateral triangle into a square, 55
   Pythagorean theorem, 66
   rectangle into squares, 62
   regular octagon into a square, 57
   square into squares, 62
Distinct magic squares, 30
Divine Proportion, a Study in Mathematical Beauty, The
   (H. E. Huntley), 1081
Doctor's Dilemma, The (G. B. Shaw), 264
Doubly-even magic squares, 41
Doubly-magic squares, 42
Douglas, J., 84, 231
Doyle's *The Adventure of the Three Students*, 360
Dreyfus, A., 177
Dudeney, A. (Mrs. H. E. Dudeney), 57
Dudeney, H. E., 43, 55, 56, 57
   Puzzles and Curious Problems, 57
Duhem, P. M. M., 359
Dunraven, Countess of, 137
Dunraven, Earl of, 137
Dunsink Observatory, 135, 138
Dürer, A., 1, 77, 1089
   approximate constructions, 87
   magic square, 35, 41
   Melencolia, 35

Eberlein, G., 209
Echols, W. H., 148

École Polytechnique, 161, 171
Eddington, Sir. A., 114*p*, 126
  Space, Time, and Gravitation, 126
Edinburgh Eleven, Pencil Portraits from College Life, An
  (J. M. Barrie), 114*p*.
Edison, T., 355
Educational Times, 123
Eells, W. C., 315
Egorov, K. N., 84
Egyptian hieroglyphic numerals, 64
Eight queens problem, 71
Einstein, A., 76, 91*p*, 96, 97, 98, 216, 222, 259, 263,
  1083
Einstein, Mrs. A., 98
Electromagnetic telegraph, 210
Elements (Euclid), 86
Elements of Natural Philosophy, The (W. Thomson
  and P. G. Tait), 145
Ellipsis, 13
English Dictionary (Bailey), 15
Equivalence, 324
Erdös, P., 113
Estes, R. A., 332
Euclidean foundation of geometry, 73
Euclid's *Elements*, 86
Euler, L., 46, 47, 51, 89, 166, 167, 226, 317, 1089
Eulerian squares, 46
  origin of, 47
Eves, H. W.
  In Mathematical Circles, 44, 84, 106, 125, 134*p*,
    144*p*, 163, 166, 170, 172, 174*p*, 183, 240, 308,
    314, 315, 348, 349, 353, 1081*p*, 1083
  Mathematical Circles Revisited, 70, 89*n*, 134*p*, 144*p*,
    230, 277, 345, 1081*p*
  Mathematical Circles Squared, 1081*p*
Extraordinary professor, 271

Fadiman, C., 92, 92*n*, 134
  The Mathematical Magpie, 92*n*
False-coin problems, 312, 312*n*
Faraday-Maxwell vector concept, 73
Fermat's last "theorem," 229, 236, 292
Fibonacci, 9
  magic squares, 44
  sequence, 44
Finger numbers, 1, 2, 3, 4, 5, 6, 7, 8, 9
  attitude for prayer, 7
  falconry, 8
  holding the bowstring, 6
  Roman versus Arabic custom, 4
Finkel, B. F., 109
Finkel's Mathematical Solution Book (B. F. Finkel), 109
First International Congress of Mathematicians

(Paris, 1900), 238, 292
Formalists, 259
Fourier series, 231
Fourth dimension, 126
Franklin, B., 30*p*
Frazer, P., 1081
Frederick II, 8
  On the Art of Hunting with Birds, 8
Frederick William III, 199
Fredholm theory, 1087
French Academy of Sciences, 181, 219, 263, 299
Frend, S. E., 118
Frend, W., 114*p*, 118
  Principles of Algebra, 118
Friedrich, A., 200
Friedrichs, K. O., 358
Frog and mouse battle, 259
Frost, A. H., 114*p*, 124
Frost, P., 124
Fuchs, K., 288
Fuchs, L., 251, 296*p*, 303
Functional dependence, 327

Galileo, 240
Gall, F. J., 354
Gauss, C. F., 158*p*, 167, 188*p*, 188, 189, 190, 191,
  192, 193, 194, 195, 196, 197, 198, 199, 200, 201,
  202, 203, 204, 205, 206, 207, 208, 209, 210, 211,
  212*p*, 216, 226, 317, 348
  Disquisitiones arithmeticae, 202
  motto, 167
Gauss, P. S. M. E., 194, 196
Gaussian foundation of geometry, 73
Gauss-Weber monument, 210
Gelfond, A. O., 292
Geometry, 63, 73, 74, 75, 76, 77, 78, 79, 80, 81, 82,
  83, 84, 85, 86, 87, 88, 89, 90, 159, 181*p*, 187,
  207, 221, 254, 309, 331, 360, 1081
Geometry of paths (O. Veblen), 131
Gibbs phenomenon, 231
Glaisher, J. W. L., 1089
Gnu, Mr. and Mrs., 329
Goethe, J. W. von, 180
  Maximen und Reflexionen, 180
Gordon, P., 226
Gosset, T., 71
Gosset, W. S., 114*p*, 127
Göttingen Mathematics Club, 243, 246, 251, 259,
  277
Göttingen Scientific Society, 236
Göttingen University, 188*p*, 190, 212*p*, 216, 225,
  271*p*
Graf Zeppelin, 89*n*

Grassmann, H. G., 296*p*, 297
  Ausdehnungslehre, 297
Greek alphabetic numerals, 64
Greenwood Publications, Inc., 26
Greer, R., 325
Grommer, J., 295
Guyaelf, 344

Habilitationsschrift (G. B. Riemann), 216
Hachette, J. N. P., 181
Hadamard, J., 177, 178, 351, 359, 1087
  determinant theorem, 1087
  *The Psychology of Invention in the Mathematical
    Field*, 348*p*
Halmos, P., 114*p*, 127
Hamilton, Sir W. R., 134*p*, 134, 135, 136, 137, 138,
    139, 140, 144*p*, 170, 349
  "On College Ambition," 134
Hankel, H., 1089
Hardy, G. H., 20, 114*p*, 129, 130, 227, 230, 232,
    307, 342, 1089
  A Course in Pure Mathematics, 129
Hartzer, F., 210
Harvard University, 91*p*, 91, 264
Hays, A. G., 335
Heath, R. V., 42, 45
Heawood, P. J., 58
Hebrew numerals, 64
Hecke, E., 235
Hedrick, E. R., 327
Heifitz, J., 317
Heisenberg, W., 104, 249, 288
Heliograph, 208
Heliotrope, 208
Hermannus Contractus, 336
Hermann the Lame, 336
Hermite, C., 177, 355, 1087
Herodian, 12
Herodianic numerals, 12, 64
Herodotus' *Historia*, 18
Hilbert (C. Reid), 234*p*, 237
Hilbert, D., 223, 224, 234*p*, 234, 235, 236, 237, 238,
    239, 240, 241, 242, 243, 244, 245, 246, 247, 248,
    249, 250, 251, 252, 253, 254, 255, 256, 257, 258,
    259, 260, 261, 262, 263, 264, 265, 266, 267, 268,
    269, 270, 276, 277, 281, 283, 286, 287, 292, 293,
    295
Hilbert, F., 255, 156
Hilbert, K. (Mrs. D. Hilbert), 223, 235, 241, 251,
    256, 269
Hilbert space, 254
Hilbert Strasse, 269
Hindu-Arabic numerals, 9, 15, 16, 17, 29, 64

Hipparchus, 140
Historia (Herodotus), 18
History of his Life and Times (W. Lilly), 172
Hitler, A., 232, 282
Hobbes, T., 14
  The Leviathan, 14
Hobson, E. W., 1089
Hohenhagen Tower, 209
Holmboë, B. M., 183
Holmes, S., 360
Hough, R., 330
Humboldt, A. von, 199, 200
Huntley, H. E., 1081, 1082
  *The Divine Proportion, a Study in Mathematical
    Beauty*, 1081
Huxley, T. H., 141

IACOBACCI, R., 285
I Am a Mathematician (N. Wiener), 100, 133
Identity, 324
I-king, 10, 31
In Mathematical Circles (H. W. Eves), 44, 84, 106,
    125, 134*p*, 144*p*, 163, 166, 170, 172, 174*p*, 183,
    240, 308, 314, 315, 348, 349, 1081*p*, 1083
Intuitionists, 259, 260
Invariantentheorie (R. Weitzenböck), 347

JACOBI, C. G. J., 163, 181
Jacobi, M. H., 163
Janus, 3
Jehovah, 32
Jenkin, F., 146
Johns Hopkins University, 102
John the Baptist, 131
Jordan, C., 174*p*, 176, 184
Journal of the American Medical Association, 264
Jupiter, 33

KAGAN, V. F., 84
Kaiser Wilhelm Institute, 263
Kant, I., 89
Kasner, E., 91*p*, 92, 92*n*, 93, 94, 95
Kästner, A. G., 226
Kazan University, 185
Kellogg, O. D., 264
Kelly, L. M., 112
Kepler, J., 359
Kerékjártó, B. M. de, 345
  Vorlesungen über Topologie, 345
Khalid, 5
Kimball, S. H., 91*p*, 108
Kingdon, F., 122
Klein, F., 184, 220, 251, 263, 269, 271*p*, 272, 273,
    274, 275, 276, 307

Index

Koebe, P., 296*p*, 305
Königsberg bridge problem, 89
Kronecker, L., 296*p*, 302
Kummer, E. E., 296*p*, 298, 299, 300, 301, 1089

Lagrange, J. L., 317, 1089
Lanczos, C., 320
  Space through the Ages, 79, 80, 83, 217, 218
Landau, E., 212*p*, 225, 226, 227, 228, 229, 230, 231, 232, 233, 291
Landau, Mrs. E., 233, 264
Large numbers, 19, 20
Latin squares, 46
Law of excluded middle, 260
Leaper, 53
Least natural number principle, 23
Legendre, A. M., 181
  Théorie des Nombres, 214
Lehmer, D. N., 24
Lemniscate, 360
L'Enseignement Mathématique, 348*p*, 356, 357
Levenspiel, O., 337
Leviathan, The (T. Hobbes), 14
Levi-Civita, T., 316
Liang I, 10
Lichtenstein, L., 100
Lie, M. S., 181*p*, 184, 1089
Lietzmann, W., 251
Lilly's *History of his Life and Times*, 172
Littlewood, J. E., 19, 20, 114*p*, 128, 129, 130, 131, 132, 133, 230, 309, 310, 339, 341, 342, 343
Lobachevsky, N. I., 181*p*, 185, 186, 187
Lockyer, Sir J. N., 141
Lo-shu, 31, 32, 40
  artistic applications of, 48
Loubère, A. de la, 40*n*
Loubère, S. de la, 40
  rule, 40
Louis XIV, 40
Loxodrome, 78
Lübsen, H. B., 201
Luther, M., 226
Lyons, L. V., 56

Macfarlane, A., 114*p*, 121, 141
Magic constant, 30, 45
Magic cubes, 45
  magic constant, 45
  normal, 45
  pandiagonal, 45
Magic squares, 30*p*, 30, 31, 32, 33, 34, 35, 36, 37, 38, 39, 40, 41, 42, 43, 44, 46, 48, 49, 50, 52, 54
  diabolic, 30
  distinct, 30

doubly-even, 41
doubly-magic, 42
Dürer's, 35, 41
Fibonacci, 44
lo-shu, 31, 32, 40, 48
magic constant, 30
nasik, 30
normal, 30
odd, 40
pandiagonal, 30, 34, 36, 37, 38, 42
perfect, 30
prime, 43
semimagic, 30, 46, 50
symmetric, 30
trebly-magic, 42

Magic traces, 49
Mairhuber, J. C., 233
Mao Tse-tung, 226
Maps, 58, 59, 61, 78, 78*n*
Marischal College, 141
Mars, 33
Mascheroni, L., 134
Mathematical bump, 354
Mathematical Circles Revisited (H. W. Eves), 70, 89*n*, 134*p*, 144*p*, 230, 277, 345, 353, 1081*p*
Mathematical Circles Squared (H. W. Eves), 1081*p*
Mathematical existence, 248
Mathematical Gazette, The, 19, 344
Mathematical Magpie, The (C. Fadiman), 92*n*
Mathematical taste, 1085, 1086
Mathematics Teacher, The, 315
Mathematische Annalen, 237, 259, 293, 316
Matthew, 1
Maximen und Reflexionen (J. W. von Goethe), 180
Mayan numerals, 64
Mayor's Audits, 16
McLoed, Mrs., 122
Medicus, 2
Melencolia (A. Dürer), 35
Memoir on a general property of a very extensive class of transcendental functions (N. Abel), 181
Mental Physiology (Carpenter), 193
Mercator, G., 78
  projection, 78, 78*n*
Michaelangelo, 1089
MICR process, 25
Miller, Mr., 123
Miller, G. A., 346
Milne, W. P., 344
  "Noether's canonical curves," 344
Minkowski, F., 219
Minkowski, H., 212*p*, 219, 220, 221, 222, 223, 224, 251, 270, 276, 277, 293, 306

INDEX

Minot, G. R., 264
M. I. T., 91*p*, 104, 107, 333
MMM (My Mathematical Museum), 89*n*
Möbius strip, 75
Modern numerals, 25
Monge, G., 207
Moore, R. L., 91*p*, 99
More Lives Than One (C. F. Bragdon), 49
Morón, Z., 62
Mosaics, 56, 66, 67, 68, 69
Moser, L., 23
Muncey, J. N., 43
Murnaghan, F., 102

Nachrichten, 262
Napier, J., 172
Napoleon Bonaparte, 174*p*, 174, 175
Nasik magic squares, 30
National Academy of Sciences, 139
Nature, 141
NBC Nightly News television broadcast, 335
Neumann, F. E., 296*p*, 296
Neumann, J. von, 254
Newton, I., 114*p*, 125, 136, 174, 183, 315
    Principia, 136
Newton particle concept, 73
New York Times, 107
Noether, E., 284, 285, 286
"Noether's canonical curves" (W. P. Milne), 344
Non-Euclidean geometry, 181*p*, 187
Normal magic squares, 30
Numbers (*See* Finger numbers; Numerals)
    large, 19, 20
    perfect, 2, 21
    prime, 24
Numerals
    Attic, 12
    Babylonian cuneiform, 64
    binary, 11
    Egyptian hieroglyphic, 64
    Greek alphabetic, 64
    Hebrew, 64
    Herodianic, 12, 64
    Hindu-Arabic, 9, 15, 16, 17, 29, 64
    Mayan, 64
    modern, 25
    Pa-kua, 10
    Persian, 64
    Roman, 16, 17, 64
    scientific Chinese, 64
    zero, 13, 17

OAKLAND University, 27
Octagon puzzle, 57

Odd-order magic squares, 40
"On College Ambition" (Sir W. R. Hamilton), 134
On the Art of Hunting with Birds (Frederick II), 8
Ordinary professor, 271
Osgood, W. F., 91*p*, 91
Oxford Group, 20
Oxford University, 116

PA-KUA, 10
Pandiagonal magic squares, 30, 34
    doubly-magic, 42
    on a hypercube, 38
    on a plane, 37
    on a torus, 36
Parable of the Sower, 1
Paris Academy (*See* French Academy of Sciences)
Parker, E. T., 46
Pascal's mystic hexagram theorem, 63
Passano, L. M., 333
Pentominoes, 72
Perfect magic squares, 30
Perfect numbers, 2
Persian numerals, 64
Pi Mu Epsilon Journal, 22
Plato, 82
    The Republic, 76
Plattner Story, The (H. G. Wells), 75*n*
Pliny the Elder, 3
Poincaré, A., 162
Poincaré, H., 158*p*, 158, 159, 160, 161, 162, 163,
    164, 165, 166, 167, 168, 169, 170, 171, 172, 173,
    350
Poincaré, L., 162
Poincaré, R., 162, 163, 165
Practical versus esthetical, 1083
Preparation-incubation-illumination, 350, 358
Prime magic squares, 43
Prime numbers, 24
Principia (I. Newton), 136
Principle of duality of plane projective geometry, 80
Principles of Algebra (W. Frend), 118
Pringsheim, A., 1089
Problems, 71, 89, 109, 110, 111, 112, 113, 113*n*,
    123, 312, 312*n*
Proceedings of the London Mathematical Society, 344
Projective geometry, 1081
Psychology of Invention in the Mathematical Field, The
    (J. Hadamard), 348*p*
Ptolemaic system, 328
Ptolemy, C., 140
Puzzles and Curious Problems (H. E. Dudeney), 57
Pythagorean theorem, 66

Q. E. D., 90

INDEX

Quaternions, 349
Quilts, 71, 72

RAMANUJAN, S., 130, 334
Raphael (Raffaello Sanzio), 1089
Ray's *Third Part Arithmetic*, 109
Reading Mercury, Berks County Paper, 115
Recreational Mathematics Magazine, 43
Re-entrant king's path, 54
Re-entrant knight's path, 50
    half-board solutions, 51
    magic half-board solutions, 52
Re-entrant strong-knight paths, 53
Reid's *Hilbert*, 234p, 237
Republic, The (Plato), 76
Ribbentrop, G. J., 190, 191
Ricci, G., 316
Riemann, G. B., 212p, 212, 213, 214, 215, 216, 217,
    218, 226, 1089
    Habilitationsschrift, 216
    hypothesis, 253, 292
Riemannian sphere, 75
Ring finger, 2
Ripley's *Believe It or Not, 15th series*, 325
Roget, P. M., 50, 51
Roman numerals, 16, 17, 64
Rosenblueth, A., 102
Rosse, Lord, 138
Rosser, J. B., 30
Royal-Dawson, F. G., 360n
Royal Irish Academy, 349
Royal Society, 117
Rudio, F., 1089
Runge, C., 276, 277
Russell, B., 86, 162, 1089
    Autobiography, 86

SAINT Jerome, 1
Sanskrit, 13
Sayles, H. A., 43
Schlegel, V., 60
    diagrams, 60
Schoenberg, I., 233
School Visitor, The, 109
Schots, M. H., 42
Schweitzer, A., 226
Science fiction, 75, 75n
Scopes, J. T., 335
    trial, 335
Scott, D. S., 72
Second International Congress of Mathematicians
    (Bologna, 1928), 266, 267, 268
Semimagic squares, 46, 50
Set theory, 340, 341

Shaw's *The Doctor's Dilemma*, 264
Shields, F. M., 111
Shrikhande, S. S., 46
Shuldham, C. D., 43
Siegel, C. L., 291, 292, 293
Sioux, 196
Smith, C. A. B., 62
Smith, H. J. S., 134p, 141, 142, 143, 219
Space method, 315
Space through the Ages (C. Lanczos), 79, 80, 83, 217,
    218
Space, Time, and Gravitation (Sir A. Eddington), 126
Sprague, R., 62
Sputnik I, 332
Steiner, J., 352
Stevenson, R. L., 146
Stone, A. H., 62
Struik, D. J., 84, 91, 97, 101, 104, 105, 106, 174, 175,
    226, 231, 249, 250, 286, 333, 345, 346, 347
    A Source Book in Mathematics, 337p
Student, 127
Student Christian Movement, 1082
Student's First Glimpse of Hades, The, 145
Sunya, 13
Sunya-bindu, 13
Swinford, L., 150
Sylvester, J. J., 114p, 119, 172, 303
Symbolic method, 114
Symmetric magic squares, 30
Synopsis of Mathematics (Carr), 334
Szekeres, E., 113
Szekeres, G., 113, 113n
Sz' Siang, 10

T and T, or *T and T'* (W. Thomson and P. G. Tait),
    145
Tait, P. G., 144p, 144, 145, 146, 147, 148, 149, 150
    T and T, or *T and T'*, 145
    The Elements of Natural Philosophy, 145
    The Student's First Glimpse of Hades, 145
Tarry, G., 46
Teipel, J. H., 192
Tender embrace, the, 1
Tensors, 74
Tesseract, 75
Théorie des Nombres (A. M. Legendre), 214
Theory of numbers, 219, 220, 229, 250, 291
Third Part Arithmetic (Ray), 109
Thomson, W., 145
    T and T, or *T and T'*, 145
    The Elements of Natural Philosophy, 145
    The Student's First Glimpse of Hades, 145
Thought curve, 360

INDEX

Tietze, H., 59
Tiling (*See* Mosaics)
Times, The, 153
Todhunter, I., 114*p*, 120, 121
Toole, J. W., 179
Topology, 75, 88, 89, 221
Travers, J., 57
Trebly-magic squares, 42
Trinity College, 133, 134, 136
Tucker, A. W., 61
Tucker, R., 114*p*, 125
Tutte, W. T., 62

UMBRAL calculus, 114
University of Berlin, 251, 296*p*
University of Dublin, 135
University of Edinburgh, 117, 144*p*, 145, 148, 152
University of Helmstedt, 202
University of Illinois, 179
University of Königsberg, 296*p*
University of Maine, 91*p*, 108, 150, 186
University of Missouri, 112
University of Pennsylvania, 233
University of Virginia, 148
UPI, 27

VAN der Waerden, B. L., 259, 261, 294
Veblen, O., 30*p*, 114*p*, 131
    Geometry of paths, 131
Venus, 33, 48
Von Neumann, J., 254
Vorlesungen über Topologie (B. M. de Kerékjártó), 345
Vulgate version, 1

WALKER, R. J., 30
Washington, G., 226
Watts, I., 118
Weber, W., 210
Wells' *The Plattner Story*, 75*n*
Weierstrass, K., 1089
Weitzenböck, R., 347
    Invariantentheorie, 347
White, W. F., 1089
Whitehead, A. N., 314
Wiener, N., 91*p*, 100, 101, 102, 103, 104, 105, 106, 107, 114*p*, 133, 178, 243, 274, 321
    Cybernetics, 105
    I Am a Mathematician, 100, 133
Wilhelm, Kaiser, 261
Willcocks, T. H., 62
Wolfskehl, P., 236
Wolfskehl Prize, 236
Wolverad, Count, 336
Wordsworth, W., 135
Wordsworth Walk, 135

XERXES, 18

YANG, 10
Ying, 10
Young, J. W. A., 1089
Yu, Emperor, 31

ZENO, 315
Zermelo, E., 278, 279
Zero, 13, 17
Zerr, G. M. B., 110
Zeuxis, 77